£14-95

Discovering Landscape in ENGLAND & WALES

TITLES OF RELATED INTEREST

Unwin Landscape Guides
Edited and written by Brian Knapp & John Whittow

1. *The Lakes*
2. *Dartmoor & South Devon*
3. *The North York Moors & Coast*
4. *Snowdonia's landscape*
5. *Britain's landscape from the road*
6. *Europe's landscape from the road*

Rocks & Fossils series
Edited by J. A. G. Thomas

1. *Snowdonia*
   M. F. Howells, B. E. Leveridge & A. J. Reedman
2. *The Weald*
   Wes Gibbons
3. *The Lake District*
   The Cumberland Geological Society
4. *The Peak District*
   I. M. Simpson

Discovering Landscape in ENGLAND & WALES

Andrew Goudie
Rita Gardner

London
GEORGE ALLEN & UNWIN
Boston   Sydney

George Allen & Unwin (Publishers) Ltd,
40 Museum Street, London WC1A 1LU, UK

George Allen & Unwin (Publishers) Ltd,
Park Lane, Hemel Hempstead, Herts HP2 4TE, UK

Allen & Unwin Inc.,
9 Winchester Terrace, Winchester, Mass. 01890, USA

George Allen & Unwin Australia Pty Ltd,
8 Napier Street, North Sydney, NSW 2060, Australia

First published in 1985

British Library Cataloguing in Publication Data

Goudie, Andrew
    Discovering landscape in England and Wales.
1. Landforms——England
I. Title   II. Gardner, Rita
551.4′0942   GB436.G7
ISBN 0–04–551076–8

Library of Congress Cataloging in Publication Data

Goudie, Andrew
    Discovering landscape in England & Wales.
Bibliography: p.
Includes index.
1. Landforms——England.   2. Landforms——Wales.
I. Gardner, Rita, 1955–   II. Title.   III. Title:
Discovering landscape in England and Wales.
GB436.G7G68 1985   551.4′0942   84-24421
ISBN 0–04–551076–8 (alk. paper)

Set in 10 point Bembo by V & M Graphics Ltd, Aylesbury, Bucks
and printed in Great Britain by Richard Clay (The Chaucer Press) Ltd,
Bungay, Suffolk

Preface and Acknowledgements

We have derived tremendous pleasure from visiting every one of the sites described in this book, sites selected because they are the best of their type. We have tried to give a wide geographical spread, though inevitably some lowland and urbanised areas are not represented. Limitations on length meant that we had to be highly selective, and so some worthy sites had to be omitted. A list of sites of geomorphological importance prepared for the British Geomorphological Research Group (BGRG) by Professor J. C. Pugh and the *Bibliography of British geomorphology*, edited for the same group by Professor K. M. Clayton, were invaluable aids in the preparation of this book. We are also grateful to the many geomorphologists whose published works we have consulted, and to leaders of various university, BGRG and Quaternary Research Association field trips who have introduced us to some of the sites.

Among those who have given direct assistance we should like to mention Denys Brunsden, Mick Day, Clifford Embleton, Mike Hart, John Norbury, Nina Piggott, Geoff Poole, Charles Sperling, David Stoddart, Marjorie Sweeting, Andrew Tatham, David Veitch, Richard West and Rendel Williams. Among those whose ideas we have found particularly valuable for individual sites, and who are not acknowledged in the above list, are Ken Addison (Nant Ffrancon), George de Boer (Spurn Head), Alan Carr (Chesil Beach), Ken Gregory (Newtondale), Brian John (Green Bridge), David Jones (Stonebarrow), Mike Kerney, Eric Brown and Tony Chandler (Devil's Kneadingtrough), John Small, Mike Clarke and John Lewin (Piggledene and Lockeridge), Bruce Sparks (Walton Common), Alfred Steers (Scolt Head), Nick Stephens, Derek Mottershead and Dave Dalziell (Valley of the Rocks), Trevor Thomas (Heads of the Valleys) and Steve Trudgill and Dingle Smith (Cheddar Gorge).

Mapping based upon Ordnance Survey maps, with the permission of the Controller of Her Majesty's Stationery Office; Crown copyright reserved.

The authors wish to thank the following for permission to reproduce photographs: The Director of Aerial Photography, University of Cambridge (p. 9, sites 10, 31 & 54); J. Boardman (site 2); Aerofilms Limited (sites 4, 7, 34 & 47); S. R. Atkinson (site 5); North Yorkshire Moors National Park (site 9); Studio Jon (site 26); British Geological Survey (sites 27 & 39); D. Brunsden (site 55). All other contemporary photographs were supplied by the authors. The following are thanked for permission to reproduce line drawings: geological map on page 5 reproduced from *Geology and scenery in England and Wales* (Trueman) by permission of Penguin Books Ltd; Nant Ffrancon diagram in site 18 reproduced by permission of the Editor, *Geographical Magazine*; marsh development diagrams in site 32 reproduced from *The green planet* (Moore) by permission of Cambridge University Press.

A.S.G.
R.A.M.G.

Contents

CONTENTS

A WONDERFUL LANDSCAPE

Have you ever wondered why there is such a variety of natural landscapes within our small island? Why are there mountains in Wales and north-west England, moorlands in Devon and Cornwall, and rolling downlands in south-east England? Ever been puzzled by those peculiar hummocks in the ground over there, or by the waterfall suddenly revealed by a bend in the road? How many people enjoy a quiet stroll by a river or an energetic ramble through beautiful countryside, and yet wish they could understand how and when it was fashioned?

The natural landscape has played a major role in human settlement of our island, controlling, in the early days, the routeways, defensive positions, and location of rich farmland. Today, with man's ability to master his surroundings, the position is reversed and the natural landscape is being modified by man, often to its detriment. However, the remaining large tracts of unspoiled countryside still offer a greater variety of scenery than almost any other area of comparable size on the face of the Earth. Thanks to local conservation groups, the National Trust, the Nature Conservancy Council, the national parks, and many of our farmers, these natural features are preserved despite the demands made upon them.

In this book we discover and try to explain some of the most appealing features ('landforms') of the natural landscape. The selected examples represent most of the major types of landform found in England and Wales. Many are visually spectacular, some are surprising, whereas others are more subdued in appearance but none the less interesting and intriguing. The reader may feel challenged to identify similar features elsewhere in Britain, using the sites described in this book as an introduction to the landscape.

Although it often appears otherwise, our landscape is transitory, being merely one stage in the grand cycle of destruction and creation that characterises the surface of the Earth. Molten material escapes from the interior of the Earth, to the surface, where it cools and hardens. Then it is slowly worn away by the elements – ice, water and wind. The rocks gradually disintegrate and decompose, and are washed back into the sea by rivers draining the land surface. Eventually these muds and sands will be forced up out of the sea to form mountain chains and hills, which will then, slowly, be worn away again. From time to time molten material continues to escape along lines of weakness, creating spectacular eruptions as at Mount St Helens in America in 1981.

The landscape in England and Wales is at various stages in this cycle. Some features, such as the mountains, are merely remnants of their former glory, of mountain chains that millions of years ago rivalled the Alps. The fundamental features – mountains, hills and vales – are usually closely related to the types of rock beneath the surface. Stronger and harder rocks are less easily worn away, tending to form the highlands, and weaker rocks form the vales. We must remember that time is also important, for no matter how strong the rock is, it cannot last forever. The wearing away of rocks by ice, wind and water, has left distinct impressions on the basic rock-controlled skeleton of Britain. It is these impressions that furnish the superb variety of scenery, and that form the basis of this book.

The raw material

Most of the spectacular landscapes of England and Wales are found within the rugged, higher lands of the north, west and south-west. Why is there such a striking difference between the areas of mountains and moorlands, and the gently undulating lowlands characteristic of the south and east? The answer lies in the nature of the rocks and of their susceptibility to being worn away by the elements.

There are three major types of rock. Igneous rocks are formed by the cooling of molten material ('magma') that escapes to the surface layers of the Earth from the interior. Metamorphic rocks are formed by baking and deformation at high temperatures and pressures. Sedimentary rocks are formed under much quieter conditions, by the accumulation of material on the sea floor or land surface. Most of the igneous and metamorphic rocks are hard and compact, and are therefore more resistant to attack at the Earth's surface than the weaker sedimentary rocks. And so it is hardly surprising to discover that many of the upstanding areas in England and Wales are composed of igneous and metamorphic rocks – the moorlands of the South-West Peninsula (except Exmoor), the highlands of North Wales, and the Lake District.

Let's now fill in some of the details by looking at the different rock types to see how they are formed, how resistant they are and what sort of landscapes they give rise to in England and Wales.

When the Earth came into being nearly 5000 million years ago it was made up of molten magma. This gradually cooled and solidified at the surface, forming a crust, but the interior of the Earth remained hot and molten. Today, when this molten material is pushed up under great pressure through cracks in the

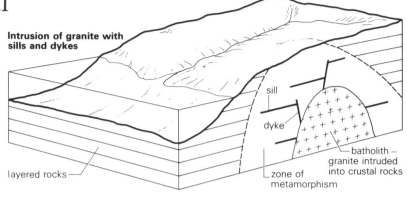

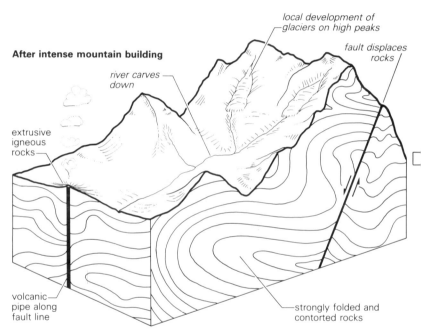

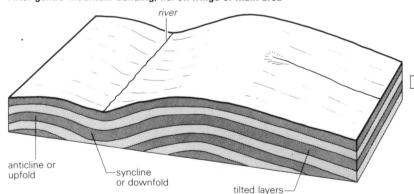

The formation of igneous, metamorphic and sedimentary rocks and the effects of mountain building

2

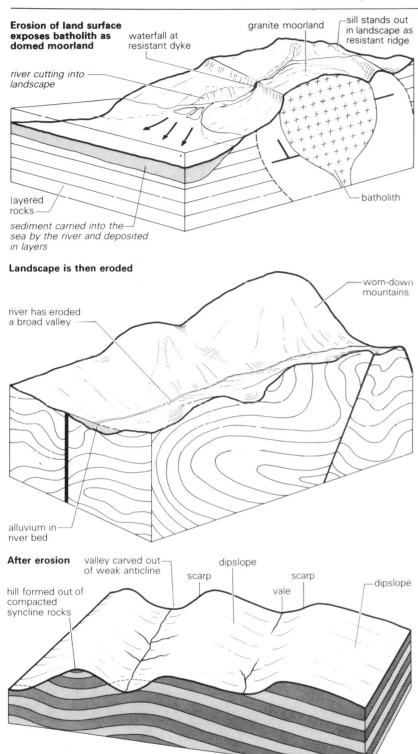

Erosion of land surface exposes batholith as domed moorland

granite moorland

sill stands out in landscape as resistant ridge

waterfall at resistant dyke

river cutting into landscape

layered rocks

sediment carried into the sea by the river and deposited in layers

batholith

Landscape is then eroded

worn-down mountains

river has eroded a broad valley

alluvium in river bed

After erosion

valley carved out of weak anticline

dipslope

scarp

scarp

vale

dipslope

hill formed out of compacted syncline rocks

crust, volcanoes are formed. On reaching the air, the magma cools rapidly to form hard, compact rocks such as basalt. The exact type of volcanic rock formed depends on the chemistry of the magma. Some volcanic eruptions also produce much softer ash, which is nothing more than the shattered rock fragments that originally blocked the volcanic pipe.

Although there are no active volcanoes in Britain today, some of our older rocks were formed in this way. Particularly well known are the volcanic rocks and ash of the Lake District and North Wales, both of which were formed in a major phase of volcanic activity about 500 million years ago, during the Ordovician period. Over 20% of the rocks that make up Britain's land surface are volcanic in origin.

Sometimes magma does not quite reach the Earth's surface and remains trapped in the upper layers of the crust where it cools and solidifies slowly. This is because the temperatures in the crust are much higher than at the surface. Such slow cooling means the crystals have a longer time in which to grow, and thus they tend to be large, as in granites.

The trapped masses of magma often form dome-like chambers called 'batholiths'. The smaller offshoots from one of these underlie the granite moorlands of Devon and Cornwall (sites 60 and 63). The magma was pushed into the crust in these areas about 300 million years ago, during the Permian period, at about the same time as reptiles were evolving elsewhere. Over the millions of years since then, the rocks in the crust overlying the batholith complex have been worn away, exposing the granites at the surface. Since these granites are hard and therefore resistant, they have been eroded more slowly than the surrounding sedimentary rocks, so forming the upstanding moorlands. However, some weaknesses and cracks do exist in the granites, and these in turn help to explain the detailed features of the landscape, such as tors (sites 60 and 63). The cracks were created by contraction as the magma cooled, and by the release of pressure on the granite as the overlying crustal rocks were removed by erosion.

Smaller amounts of magma may penetrate outwards from the main chamber as 'veins' along weaknesses in the crustal rocks. Where the magma seeps between two layers of rock it solidifies to form a thin sheet, termed a 'sill'. 'Dykes' result when the magma exploits a line of weakness that cuts through different layers of rock. The igneous rocks formed as the magma cools are usually relatively resistant and after erosion they stand out in the landscape. Much of Hadrian's Wall (site 1) was built along part of a giant sill complex – the Great Whin Sill – which also gives rise to waterfalls further south (site 3).

All of the rocks mentioned above are derived from the cooling of magma and are termed 'igneous'. The volcanic ones are called 'extrusive igneous' because the magma is extruded onto the Earth's surface, whereas the others are called 'intrusive igneous' because the magma is emplaced or intruded into the Earth's crust.

When magma is forced into the crust it is red hot, and the heat bakes and deforms the surrounding rocks. The intruded mass also exerts some pressure on the rocks, squeezing and compressing them. The baking and pressure often result in partial melting and reconstitution of the surrounding crustal rocks, which are hardened and made more compact in the process. This change in the nature of the rocks is termed 'metamorphism' and a characteristic suite of metamorphic rocks is formed.

The type of metamorphic rock formed depends on the original characteristics of the rock, its nearness to the red hot magma and the amount of pressure exerted. The degree of alteration diminishes with distance from the magma intrusion. Rock types include gneiss (highly altered clays and sandstones), schist (moderately altered clays and sandstones), marble (altered limestone), quartzite (altered pure sandstone) and slate (moderately altered pure clays). Metamorphism generally renders all the rocks more resistant to break down than their unaltered counterparts, and the most highly altered are usually the most physically compact and resistant. South-west England provides a wonderful example of the close link between igneous and metamorphic rocks. The intruded magma solidified to form the granites and caused baking in the surrounding rocks. This has left more or less concentric zones of altered, metamorphised, rocks around the granites.

Sedimentary rocks make up the greater part of England and Wales. They underlie the moorlands of mid and south Wales, infill between the igneous and metamorphic rocks of the south-west, and are responsible for the typical scarp and vale landscapes of the south and east. Why, then, if they are all sedimentary rocks, are the landscapes so varied? The answer, once again, is because there is a great variety of sedimentary rocks all of which have differing resistance to the attack of the elements – water, wind and ice. Alternations of these various types give rise to alternating hilly and low-lying areas, such as between Manchester and Grimsby, or London and Gloucester.

A look at the geological map will reveal distinct bands of different types of sedimentary rocks running down the length of England. These rocks, in general, become younger towards the south-east. They mark successive stages in the development of the land surface. Thus, quite simply, the igneous and metamorphic rocks of the north and west, and the old sedimentary rocks found around them, formed the core of the island, onto which younger and younger sedimentary rocks have been grafted.

The sedimentary rocks are weaker than most of the other types because they are made up of layers of loose materials – muds, sands, gravels and shells – that have been compressed by the weight of overlying rocks or the pressures associated with earth movements, or that are held together by a cement. Thus, the grains have not been welded together by extremes of heat or pressure, as in metamorphic rocks, nor have they grown together as magma cooled.

How, then, are the various sedimentary rocks formed and why are they so diverse? The loose particles that make up the rocks come from two major sources. The first is the present land surface, for as the surface rocks are attacked by ice, wind, rain water, and sometimes even by man, they are gradually worn away, leaving a lowered landscape, and producing a lot of loose debris. Much of the debris gradually makes its way via the rivers to the sea, where it is deposited, and over time collects into layers. Depending on the nature of the eroded rock, the way in which it was broken down and the capacity of the rivers to carry the material, the particles reaching the sea could range in size from a fraction of a millimetre to huge boulders that could only be carried in a large flood. Typically mixtures of clay sizes, sand grains and pebbles are dumped on the sea floor. The clays may be compressed to form shales, like those in northern England of Carboniferous age. Loose sands may be cemented, often by lime precipitated from sea water, to give sandstones. Pebble layers are often held together by clays or lime to form conglomerates (rounded pebbles) or breccias (angular pebbles).

The second main source of loose particles is the sea itself. Shells, the

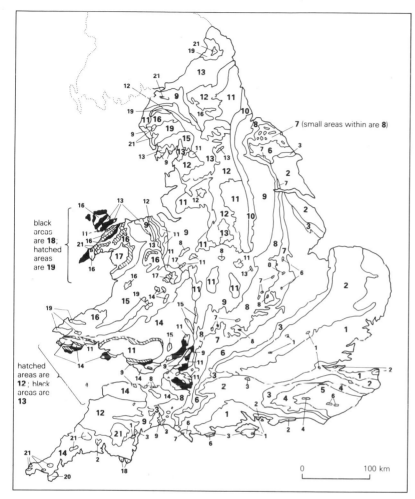

Key
1   Tertiary (with Marine Pleistocene)

Cretaceous
2   Chalk
3   Upper Greensand and Gault Clay
    Lower Greensand and Speeton Clay
4   Weald Clay
5   Hastings Beds (Wealden Sands)

Jurassic
6   Upper (clays, Corallian, Portland and Purbeck)
7   Middle (oolites and cornbrash)
8   Lower (Lias)

Permo-Triassic
9   New Red Sandstone
10 Magnesian Limestone

Carboniferous
11 Coal Measures
12 Millstone Grit and Culm Measures
13 Carboniferous Limestone

14 Devonian (Old Red Sandstone)
15 Silurian
16 Ordovician
17 Cambrian
18 Precambrian metamorphic rocks

igneous
19 volcanics
20 serpentine
21 granite

7 (small areas within are 8)

black areas are 18; hatched areas are 19

hatched areas are 12; black areas are 13

0   100 km

A geological map of England and Wales

skeletons of microscopic animals, and a lot of other organic debris, also collect on the sea bed. These are cemented together to form limestones, that is, rocks rich in calcium. Well known examples include the Cretaceous age chalk of south-east England and the Carboniferous Limestone so common in the north.

The sedimentary rocks that have been discussed so far have all formed in the sea. However, they can also form on land, either from chemical precipitation or, more commonly, when the loose particles fail to reach the sea, collect inland and are cemented together. Such a situation

often exists in deserts. As you can imagine, therefore, sedimentary rocks that formed on land are not widespread in Britain. In fact, Britain was in desert latitudes only twice in the past 600 million years, that is during the Devonian and Permian/Triassic periods. Sedimentary rocks dating to these times include the remains of cemented sand dunes and chemicals precipitated in salt lakes (site 15).

As you can see, sedimentary rocks are very variable. The sizes, hardness and chemistry of the particles making up the rocks vary enormously, as does the way in which the particles are held

together. This accounts for their different responses when attacked by the elements. Clays and shales are composed of very fine and soft particles that are easily worn away, and so they typically form vales in the landscape. Sandstones and limestones are more resistant and remain as higher areas. But, you may say, some limestones, such as chalk, are quite soft, and the lime can be dissolved by rain water, and so why do limestones form the Downs, Chilterns, Cotswolds and Peak District?

The answer is because resistance to attack is not solely to do with compactness and hardness of the rock. Limestones are permeable, that is, they allow water to pass through the rock because of the existence of pores and cracks (called 'joints'). Rain water falling on the limestones soaks in and does not often run across the surface of the rock as it would across the impermeable igneous rocks or shales. The erosion, therefore, takes place not on the surface of the rock, which remains high, but as the rainwater drains underground. Some sandstones are also permeable.

The chemistry of the minerals making up the rock, and the cement if there is one, also affect the way the rock responds to attack. Once the minerals have been broken down, the rock is more easily worn away. Sandstones, for example, are often quite resistant because they contain large amounts of tough quartz minerals.

There is one last problem concerning sedimentary rocks. How is it that these rocks now form part of the land and yet so many of them were laid down in the sea? In a few cases a fall in sea level has been responsible (site 53), but mostly the answer concerns the idea of mountain building. This is when large areas of the land masses and sea floor are crumpled and heaved up over several millions of years to form spectacular mountain chains.

Mountain building happens because the Earth's crust is divided into segments, termed 'plates', which sit on top of the molten material below. Plates vary in size and in the amount of continent and ocean on each. In all there are twelve major plates and many small ones that fit together much like a simple jigsaw puzzle. The plates, however, are not static, for they all seem to move independently, at rates of about 2 cm per year. The cause of movement is still not well understood, but the slow movement of plates around the surface of the Earth over geological time does explain why, for example, Britain has moved from a position around

the Tropics at approximately 130 million years ago, to its present position.

The movement of plates has created some of the grandest features on Earth. Where two plates slide past each other in opposite directions, vast fault systems exist, and earthquakes and volcanic activity are common along them. This is the case in the San Andreas Fault in North America. Where two plates move apart the gap is continually filled by a ridge of volcanic material that wells up from the Earth's interior, creating new crust. The ridge that divides the floor of the Atlantic Ocean is a good example. Where two plates are moving towards each other, two things can happen. First, one plate may dive beneath the other and sink back down into the Earth's molten interior, where it is re-absorbed. In this way crust is very slowly destroyed. In the second case, the two plates slowly collide and buckle under the severe pressure. This often happens if the margins of both plates have land masses on them, rather than oceans. Such a collision and the resulting upheaval over several million years created the Alps. The rocks are pushed, contorted, folded over and over, and some will fracture under the strain, forming faults. The great crustal disturbance is often accompanied in the more intense areas by widespread volcanic activity and the intrusion of magma into the crust. That is, igneous rocks are formed, and the heat from the intrusion combined with the great pressures of mountain building create, in turn, metamorphic rocks. This is why the formation of igneous and metamorphic rocks often goes hand in hand with mountain building. Thus, over hundreds of millions of years the plates rearrange themselves, and the shape and positions of the continents and oceans change.

The earth movements associated with mountain building are felt over

hundreds and thousands of kilometres, and they may raise the surrounding areas of sea floor, and the sedimentary rocks on them, well above sea level. For example, earth movements associated with the building of the Alps, some 25 million years ago, were felt in south-east England. Although far less intense than in the Alpine area itself, they were sufficient to raise the sedimentary rocks of Tertiary age above sea level, and to fold them gently at the same time.

The mountainous areas of Britain probably formed in a similar way, but it took place so long ago, during the Ordovician, Devonian and Permian periods, that the mountains have been more or less worn away by now. The folds and faults created during these episodes of mountain building are seen in rocks in many areas (for example, site 59), and they can play an important role in the development of the landscape. The line along which the rocks snap and slip under extreme pressure is called a fault. Movement along the fault will eventually bring rocks of very different character next to each other, and will also create a zone of crushed and weakened rock that is then easily eroded.

Folding involves crumpling rather than snapping. As the layers of rock become compressed they are deformed into a series of crests, called 'anticlines', and troughs, called 'synclines'. Sometimes the folding is so severe that the rocks are stood on end or even turned right over. In general, the degree of folding depends on the amount of pressure exerted and hence on the distance from the centre of mountain building. Folding affects the strength of the rocks, in much the same way as bending an eraser. Rocks in a downfold (syncline) are compressed and strengthened, whereas those in an upfold (anticline) are stretched and weakened. Erosion is easy along the weakened crest of an anticline, and so, strangely, valleys are often carved out of upfolds, and the downfolds remain as hills. Snowdon is an example of a downfold mountain. In contrast, it was an upfold in the Cretaceous rocks in the Weald that weakened them sufficiently for erosion to remove the surface rock, the chalk, revealing the older rocks that had previously lain buried beneath. A journey from London to Brighton takes you across the north and south remains of the chalk – the Downs – and the older sandstones and clays that are exposed between them.

As you will have gathered by now, Britain is a geologist's paradise. It is made up of a great variety of rocks that have been formed in different ways and in different places more or less continuously over the past 600 million years and more. The differing ages and resistance of these rocks to physical and chemical attack accounts for the broad features of the British landscape – the skeleton.

Old and tough volcanic rocks are responsible for the Lake District and higher hill masses in north-west Wales. Weaker sedimentary shales and sandstones of a similar age underlie the mid-Wales moorlands. The deep-coloured and hard Old Red Sandstone of Devonian age appears in the Black Mountains, Brecon Beacons and in south-west England. The well known Carboniferous Limestone and Millstone Grit, both sedimentary rocks, give rise to the splendours of the Peak District (sites 11, 13 & 14) and the Welsh Coalfield (site 29).

These grits and limestones were lifted well above sea level during a phase of mountain building at the end of the Carboniferous period, and during this unstable phase magma was also intruded into the crust in the South-West Peninsula, Northumberland and the Lake District. Erosion over the millions of years since has revealed the batholiths, dykes and sills composed of igneous rocks, the best known being Dartmoor, and Bodmin Moor.

Central, east and south England owes its major landscape features to the younger sedimentary rocks that were deposited in the sea adjacent to the older land mass and then uplifted and warped during the Alpine mountain building phase. The gentle folding and subsequent erosion has resulted in many different rocks being exposed at the surface. The weak clays form vales and the more resistant sandstones and limestones underlie the higher landscapes. Particularly important are the Oolitic Limestone of Jurassic age and the Greensand and Chalk of Cretaceous age. Thus, the clay vales alternate with the sandstone ridges and limestone downlands. The tilt of the rocks often encourages one edge of the downs or higher areas to be steeply sloping (the scarp slope) and the other to be much gentler (the dip slope). This is well seen in the Cotswolds where the scarp faces west over the Severn valley (site 42).

Still younger, Tertiary age, deposits of poorly consolidated sands, clays and limestones are mostly found in the London Basin and New Forest areas. The sands tend to be infertile and give rise to heathlands and forests in Dorset (site 51) and Hampshire.

And so the constant cycle of growth and decay continues. Rocks are formed, exposed at the surface, subjected to the elements, and eroded. Over an extremely long period of time the landscape would eventually be worn down to a flat and featureless plain, unless heaved, contorted and injected with new life by rare events of mountain building. What we see today is merely a passing phase in the continual evolution of the landscape, and if you could come back in 100 million years time it would be very different.

But the landscapes we see around us, although broadly controlled by rock type, are fashioned in detail by the processes of removal and dumping of material that take place on the surface. Today, many of these fashioning processes are associated with landsliding and the action of rivers, but in the recent past, during the Ice Age (the Pleistocene period) much of Britain's surface was shaped by ice. The dramatic contrast in these conditions over the past 2 million years, combined with the variety of rock types we have just discussed, have created the wonderful wealth of landscapes we see around us. It is time we looked at the processes making the landscape.

Making the landscape

The various rocks we discussed just now provide the raw material for the landscape. The resistance of these rocks to attack, taking into account the time they have been exposed and the degree to which they have been' uplifted by mountain building, has determined the skeleton of the landscape. But how are the rocks actually fashioned by the elements and how are the detailed features formed, such as the valleys, cliffs and gorges?

Sculpting of the landscape involves, first, the disintegration of the rocks, a process of weathering. The resulting rock debris is then removed, a process of erosion, and transported away from its source rock. Finally, the debris is dumped either on the land surface or in the sea, a process of deposition. Weathered debris may be eroded and deposited many times, and by different agencies, before finally reaching the sea. The ways in which weathering, erosion, transport and deposition take place are largely determined by the climate. Ice is the main agent in cold, polar and arctic, regions, whereas water is the most important agent in temperate areas such as in Britain today. Where water is scarce, as in deserts, wind plays the major role in fashioning the landscape. Let's look in more detail at the processes of weathering, erosion and deposition that commonly occur in England and Wales at the moment.

Weathering is the breakdown of rock by physical, chemical and biological means, and often leads to the formation of a soil out of the broken down residue. One of the most important causes of breakdown is physical weathering during the winter months as a result of frost. Water expands on freezing and, if this takes place in joints or pores in the rock, the pressures exerted may

be enough to fracture the rock, just as the tops of milk bottles are forced off on a very frosty morning. Frost weathering, as it is called, tends to produce sharp, angular fragments of rock which, when falling off a cliff, may accumulate at the base and gradually blanket the cliff face. The spectacular screes at Wastwater show this well (site 5).

There are many types of chemical weathering, all involving a change in the chemical constituents of the rock or soil. It is most striking in limestone areas through the action of solution, whereby the alkaline calcium carbonate that makes up the limestone is dissolved by rain water, which is usually slightly acidic. The calcium is removed in solution as the water drains through the pores and joints in the limestone, often leaving great caverns and underground channels within the rock, such as at Dovedale (site 13). It is strange to think that the high masses of limestone are often riddled with hollows, passageways and streams underground. Less spectacular, but very common, is the chemical decay of minerals in the rocks, leading to the formation of clays.

Biological weathering is the effect of the growth of trees and plants and the burrowing of animals on soil and rock. Plant roots can penetrate joints and weaknesses and gradually prise the rock apart, while earthworms can rework over 20 tonnes of soil per hectare each year, significantly weakening the soil structure.

Once the rock surface has been broken down by one or more of these three types of weathering, the material is much more easily removed by erosion. Erosion involves the picking up of available sediment, such as soil and rock debris, by some means of transport, such as water, ice or wind, or its removal under gravity. At present in Britain, most

of the erosion, transport and deposition is associated with the slipping of soil and rock in various ways (mass movement), and with river and marine activity. Wind erosion also occurs in localised areas, as in the Breckland of East Anglia (site 36) or on exposed coasts (sites 49, 58), where there are vulnerable sands or sandy soils.

Mass movements are particularly widespread in the landscape of Britain, though much passes unnoticed, for only the catastrophic examples attract attention, such as the Aberfan disaster. These catastrophic types include the collapse of cliff faces and valley sides – the collapse occurring as a fall, a slide, or a muddy flow of debris and water. In many cases the movements take place when the strength of the slope is lessened by disturbing forces such as river or marine erosion undermining the base of the slope, the seepage of large volumes of water following exceptional falls of rain, and the activities of man.

There are various types of mass movement. Rock falls involve the free movement of material from a steep slope, often because frost has levered fragments from a jointed rock. Rock slides, in which blocks of rock shear away from the slope and slide downhill, are common in Britain (site 11). The rotational types of slides, where the blocks slip along a curved surface and tilt backwards in the process, are especially common, as on the Dorset coastline (site 55). Debris flows involve the rapid movement of a slurry of water and sediment under the influence of gravity, and thus are mid-way between river flow and the drier forms of falling and sliding mentioned above. The unspectacular cousin of falls, slides and flows is soil creep. It is the slow movement, imperceptible to the eye, of material down slope under gravity. Creep is aided when particles are loosened and dislodged from the soil. This may happen when the soil moves slightly as it freezes during a frost, or when it expands and contracts on wetting and drying. The activity of worms and the trampling of animals also aids creep, which often leads to a corrugated appearance of the slope.

Once rock and soil material has been weathered and then transported down the valley side by various mass movements, it reaches the river. The river carries this load away, if it has enough energy, in one of four main ways. Soluble material, derived from the chemical weathering of rocks, is carried away in solution; fine particles are carried along in suspension, giving the muddy, brown colour to many rivers; coarser fragments are bounced along the floor of the river ('saltation'); while the coarsest fragments, if they are moved at all, are rolled along the bed (as 'bedload').

If the river still has energy left after carrying its load of sediment, it will erode material from its channel, thereby adding to its load. However, if the river moves into an area where its gradient – and hence its energy – is less, or if flooding takes place, part of the sediment load will be deposited. When a river floods, the water and contained suspended material flow onto the surrounding valley floor and invade the low-lying area adjacent to the channel, called the 'floodplain'. The water on the floodplain quickly slows down and loses the energy to carry its sediment. Therefore, the sediment settles out and gradually, over many floods, builds up the surface of the land, producing fertile soils in the process.

Marine activity is similar to river activity. Like rivers, the sea erodes, transports and deposits. Its load of sediment is collected from river mouths, cliff erosion and from the sea bed itself. Much of the fine sediment brought down by the rivers is either carried out to sea in suspension or is trapped at the shore behind barriers such as spits. Here it collects in the quiet waters to form salt marshes and mud flats. In contrast, most of the coarser material, the sand and small pebbles, is carried along the shore by the process of longshore drift. It works like this. The waves washing in obliquely to the shore carry material with them. However, the return flow of the waves, which may also carry material, is at right angles to the shore. And so with each wave material moves a small distance along the shore. Where the wave energy lessens, because the water shallows or in the lee of an obstacle for example, the load can no longer be carried and it is deposited. Features such as the famous Harlech Spit (site 22) are formed when sands are carried along the shore and are deposited part of the way across the mouth of an estuary.

These, then, are the sorts of processes that are sculpting our land surface today. They are eating it away here and building it up over there. The landscape is being fashioned before our eyes, very slowly in some places but frighteningly fast in others. But the landscape has a memory, and it still reflects other, dramatically different, processes that existed in intervals throughout the past 2 million years – the Ice Age.

Patterns of coastal deposition are well seen at Scolt Head Island (see site 32), which is a barrier island with a spit at its western end (foreground)

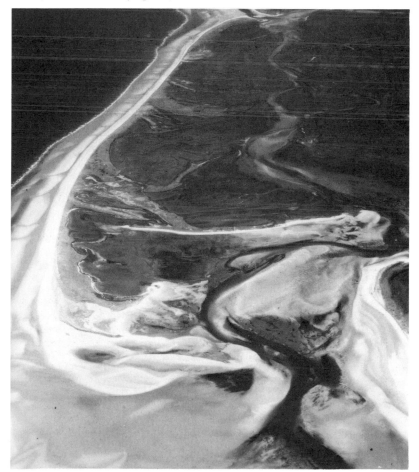

The long winter

During the Ice Age (the Pleistocene period) Britain experienced severe changes in climate, varying from times when conditions were broadly similar to those at present, to times when there were arctic conditions. In the arctic phases the highlands were covered with glaciers, while the lowlands were smothered beneath great sheets of ice, and subjected to severe frosts and a frozen subsoil. Britain would then have had a barren landscape akin to that of Alaska and Siberia today. Much of the present landscape is a sculpted relic of those icy times, for the ice has great powers of erosion. In addition, the deposits dumped from glaciers in the cold phases (called glacials) and reworked by rivers in the warmer phases (called interglacials), coat the pre-existing landscape much like icing on a cake. The valleys, lakes, corries, gorges, ridges and hummocks (sites 4, 18, 19 & 20) of the Lake District and Snowdonia are but a few of the striking reminders of the Ice Age.

The ice advanced and retreated several times, leaving a complex history recorded in the landscape. At its maximum, about 150 000 years ago, the ice sheet over Britain extended down to the north coast of Devon and Cornwall, and eastwards through Oxford to north London. Warm, interglacial, conditions prevailed between about 120 000 and 80 000 years ago, and then the ice advanced again. However, this time, even at its peak around 18 000 years ago, the ice covered little of the Midlands, and only just reached the east coast. The last 10 000 years (the Holocene) have seen a return to conditions like today, but it is likely that some thousands of years hence the ice will advance once again. Together, the Holocene and the Pleistocene periods make up the Quaternary era.

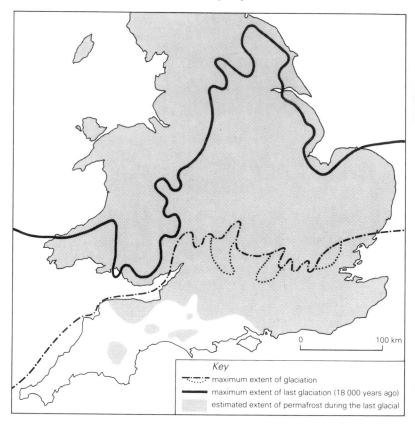

A generalised map of the extent of ice sheets and permafrost in southern Britain

Key

⋅⋅⋅⋅⋅⋅ maximum extent of glaciation

—— maximum extent of last glaciation (18 000 years ago)

▒ estimated extent of permafrost during the last glacial

0   100 km

Glacial processes of erosion and deposition can give rise to a dramatic set of landscapes. Glaciers attack the rocks in their path by freezing around them and plucking them from their surroundings; by scouring and grinding them with the coarse fragments of rock that they transport; and by eroding them with the huge quantities of water that they release on melting (meltwater). Since glaciers erode the valley sides and the entire valley floors, they tend to form broad valleys with U-shaped profiles (site 18). This shape contrasts with the typical V-shaped profiles produced by rivers as they carve downwards.

Glaciers and ice sheets deposit the material they have eroded and transported as great masses of poorly sorted gravels, sands, silts and clays, called boulder clay. These deposits generally look irregular and hummocky, although well shaped ramparts of boulder clay (called 'moraine') may be dumped at the margins of glaciers (site 19), and neat egg-shaped mounds of boulder clay (called 'drumlins') are occasionally fashioned beneath the glaciers (site 2). Often, as in north Wales, the moraines are dumped across the valley, blocking it, and ponding up the rivers to form tranquil lakes (site 21).

Huge streams flow beneath most glaciers, and in some cases are capable of carving steep, narrow gorges into the bottom of the valley (site 20). Issuing from the ends (snouts) of the glaciers and ice sheets, these streams serve to remove much debris from the base of the ice out onto the plains beyond, and hence they are called 'outwash streams'. The finer sediment is removed in suspension by the stream, but the coarser pebbles and sands cannot be carried so far and are generally deposited just beyond the glacier. Thus the debris washed out by the streams (site 33) is much better sorted than the boulder clay dumped by the ice. When the ice finally melts, the swollen streams have even more power with which to demolish the landscape.

Conditions were also severe in areas not covered by ice sheets or glaciers. The subsoil was permanently frozen, in a state called permafrost, to depths of several metres, and the landscape was barren. The permeable rocks that would normally allow water to pass through them were sealed by ice (permafrost), and so the flood water generated by springtime melting of snow and the top layers of ice was forced to flow across the surface of these rocks. This is how many of the valleys were carved into permeable rocks such as limestones (sites 8, 13 & 47). Today, water falling on these rocks sinks in and flows underground. In addition, the alternate winter freezing and summer thawing of the top layers of soil made them so unstable that when saturated by water in the summer they flowed across the landscape – a process called solifluction. The growth of ice within the soil and its subsequent melting created, in places, a confusion of hollows in the landscape, like bomb craters (site 31). Above all else, frost weathering was prevalent, giving rise to huge accumulations of shattered rock fragments beneath cliffs (sites 5 & 16) and generally breaking up the land surface.

Finally, it is worth remembering that, during glacials, much of the world's water was stored in ice caps rather than in the oceans. As a result sea levels worldwide were low, possibly as much as 100–150 metres lower than present. Large areas of the sea bed were left high and dry. The North Sea, the Irish Sea and the English Channel disappeared, leaving Britain linked to Ireland and the Continent. As the ice melted, at the end of the glacial phases, water flooded back into these areas and sea level rose. The last glacial phase ended about 12,000 years ago and the sea reached its present level approximately 6000 years ago. Sea level has been a little higher in some of the previous warm interludes ('interglacials') when more of the world's ice melted than at present (site 53).

As we have seen, there are two critical factors in the shaping of the landscape. The first is the types of rock and their resistance to attack. The second is the action of the elements (the 'processes'), and in particular the alternation of icy conditions with those similar to present. The landscape is thus the result of the interaction between rock type and processes over the time since the rocks were formed. The evidence of the recent processes is, of course, better preserved than that of the very ancient processes that took place millions of years ago. These actions or processes have, in places, broken down the rocks and eroded the debris, leaving a carved-out landscape, and elsewhere they have built up a new landscape by dumping the eroded material. Some features are a response to modern processes, whereas others are the relics of past, icy conditions. But, whatever moulds the details of the landscape, the rocks will always determine the broad characteristics of the skeleton.

As you walk around our landscape in England and Wales perhaps you can start to unravel the multitude of factors that combine to explain its present shape. What sort of rock is beneath you, and does the landscape change as you cross onto a very different type of rock nearby? Can you see, in the rocks exposed in the cliffs for example, the legacy of the mountain building that took place millions of years ago? How have the elements carved the landscape you see in front of you? Was it mainly the action of ice that created the valley, or has rainwater and the work of rivers been responsible? The rest of the book sets about discovering the wealth of landscapes in England and Wales. We explain how specific features, often the best examples of their kind, came into being, and how you can explore and understand them. Through these examples we hope you will gain greater insight into the origin and evolution of our wonderful landscapes.

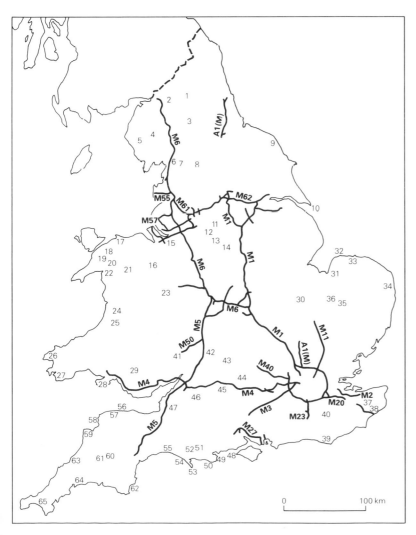

The locations of the sites

1   The Great Whin Sill

In AD 43 the Romans first invaded England. They established control over the (to them) backward English, and within 100 years the country was dotted with villas and imposing towns, all linked by a network of roads and tracks. As they tried to push northwards and westwards they met with more and more resistance from the Celts in Scotland and Wales. Marauding bands of Celts proved such a threat that the Romans felt the need to contain them, and in AD 123 set about building huge defences. Hadrian's Wall, a monument to the power and efficiency of the Romans, effectively sealed off Scotland, marking the northwestern boundary of the Roman Empire.

The wall, built of stone and turf, and originally over 6 m high in places, extended some 120 km from Bowness in Cumberland to Wallsend in Northumberland. Only parts remain today, one of the most dramatic stretches being between Peel Gap (NY 752676) and Housesteads (NY 790688), where there is an excellent excavated garrison fort. Even though large parts of the wall have crumbled over time, its route is easily seen, for the Romans chose a natural defensive site on which to build it, so providing an even more formidable barrier to invaders. Over much of its length the wall is built upon a narrow ridge of hard, dark-coloured rock called 'dolerite'. The cliff face of the dolerite beneath the wall can be seen at Housesteads, where it is over 20 m high. Just to the east of the fort, a careful look at the cliff will reveal that the dolerite has many cracks (joints). In places the pattern of joints gives rise to columns in the rock.

Dolerite is formed as molten magma from the interior of the Earth is pushed up into the Earth's crust, and then cooled and solidified. In this case the magma was pushed between layers of Carboniferous limestones, sandstones and shales to form a thin sheet of rock called a sill – the Great Whin Sill. As the intrusive magma cooled, it contracted, forming the characteristic joints in the rock. Since the intrusion, the Carboniferous rocks and the dolerite have been gently tilted.

The Whin Sill extends over a large area from Middleton in Teesdale in the south to Bamburgh in the north and is responsible for other natural features in the area (see site 3). It varies between about 30 m and 60 m in thickness, and gradually thins westwards. However, in most places the sill of dolerite is not seen on the surface of the ground, but remains covered by the rocks of Carboniferous age, the same rocks

A view along the remains of Hadrian's Wall atop the Great Whin Sill at Housesteads

into which it was pushed. It has been exposed in a few areas by erosion of the overlying rocks, especially the weak shale bands. Thus, the hard and tough dolerite, which is very resistant to erosion, stands out as ridges in the landscape. The great ridge beneath Hadrian's Wall is one of these, and is probably the most spectacular outcrop of the sill. It is like a breaking wave, sweeping up smoothly on the south side and dropping dramatically on the north.

The view from Hadrian's Wall at Housesteads encompasses rugged Carboniferous Limestone hills rising to the north and south, separated from the sill by broad valleys, and the sill itself marches up and down into the hazy distance to east and west. And when the mists descend it looms out of the landscape like a rocky Loch Ness monster. It is easy to imagine that the landscape has changed little over the centuries, and that the Romans are still guarding the northern approaches. This is excellent walking country, with many features of interest, but the best sites to view the wall and the sill are between Peel Crags and Sewing Shields, with good access to the wall and the excavated fort with its museum at Housesteads, as shown on the map. At Vindolanda, a little way to the south-west, a turret and a length of wall have been fully reconstructed.

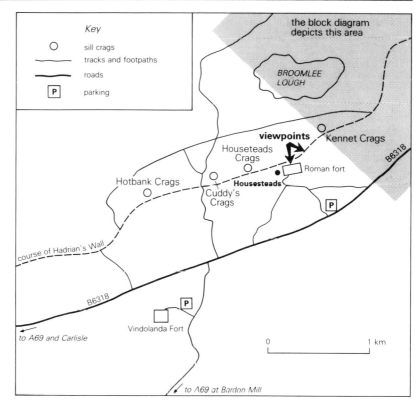

Map sheet 87

Viewpoint grid references: NY 790688 NY 793692

Facilities: interpretation centre and museum

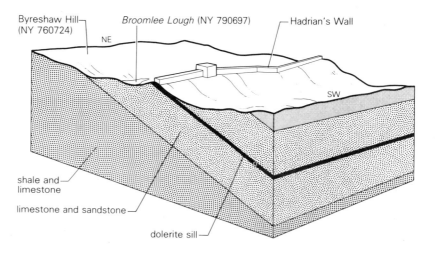

A slice through the landscape at Hadrian's Wall

2   Drumlins of the Eden Valley

Where in England do you think you can find more than 600 small oval hills, all roughly the same size and shape, in the space of 300 square kilometres? The answer is in Edendale, south of the river Eamont, in the Eden Valley of Cumbria. The hills are so regular that, viewed from above, they look like eggs in a basket, hence the term 'basket of eggs' landscape commonly used to describe such clusters. Most of them in this area have steeper southern ends, elongated tails to the north, and trend in a SE–NW direction, parallel to the valley sides. They are called drumlins, from the Gaelic word *druim*, meaning mound or rounded hill. Composed of boulder clay (a mixture of rock fragments, sands and clays dumped from glaciers during the Ice Age), they have potentially quite fertile soils, but their shape and size often makes farming rather difficult.

Thus, drumlins make up clusters of streamlined hills in some of those areas that were glaciated during the Ice Age. They were formed under moving ice, at the base of the glacier, with their longest dimension parallel to the direction of glacier flow. That explains why they are usually found parallel to the valley sides, down which the glaciers flowed. However, there is some controversy as to their origin. Some people believe that they are formed when a pre-existing sheet of glacial deposits is reshaped into mounds by ice readvancing over it. The alternative, and more likely, idea is that the drumlins are fashioned out of the boulder clay, as it is dumped around a nucleus of frozen boulder clay or rock, by moving ice exerting great pressure. The shape of the drumlin reflects these pressures. The upstream end is broad and rounded but quite steep. The ice movement blunts this end and elongates the drumlin downstream, giving a gently

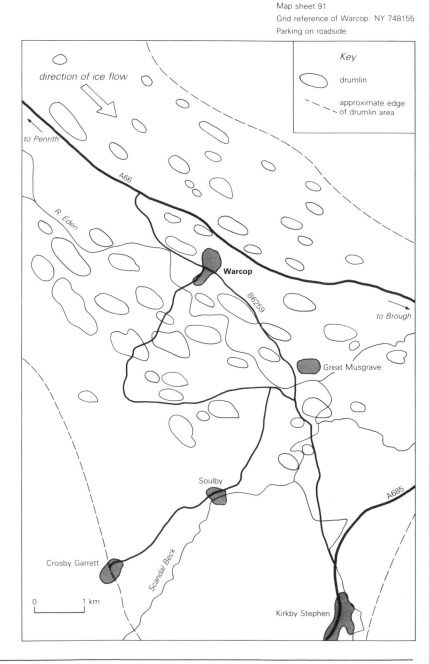

Map sheet 91
Grid reference of Warcop: NY 748155
Parking on roadside

Key

⬭ drumlin

- - - approximate edge of drumlin area

direction of ice flow

to Penrith

A66

R. Eden

Warcop

B6259

to Brough

Great Musgrave

Soulby

A685

Scandal Beck

Crosby Garrett

0   1 km

Kirkby Stephen

tapering and sloping end. Occasionally one drumlin is superimposed on another, giving rise to an irregular-shaped mound.

The drumlins in the Eden Valley are part of a much larger belt extending from south-west Scotland into the Solway area and across the Tyne Gap into north-east England. Indeed, the high ground of the Lake District and North Pennines is almost entirely surrounded by drumlins, whose trends provide evidence of local ice movements. All of the drumlins occur in valleys and lowlands rarely more than 300 m above sea level. Although these are probably the best in England, some of the most striking drumlins in the world are those that extend across a large portion of the northern half of Ireland from County Down to Sligo.

The best area for viewing drumlins in the Eden Valley is shown on the map. They certainly cannot be missed around Great Musgrave and Warcop, although half the fun is in trying to identify them in other areas too.

These elongated hills are drumlins

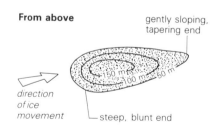

From above

gently sloping, tapering end

direction of ice movement

steep, blunt end

From the side

ice movement

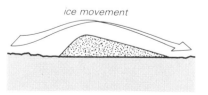

The shape of a drumlin in relation to glacier flow

3　High Force: a remarkable cataract

High Force, one of the larger
waterfalls in England, was a great
source of enjoyment to the
Victorians, who regarded it as one of
the wonders of the countryside.
Fortunately, it retains its Victorian
charm and the waterfall is
approached by a delightful stroll
down the wooded valleyside. On the
way, one passes exposed and gnarled
tree roots, clinging onto the crevices
in the rocks, which are themselves
covered with mosses and oozing
water gently down the hillside. The
smell of wild garlic is in the air and
the distant roar of the waterfall can
be heard. In the narrow gorge-like
valley below, the stream meanders in
summer through a pebble-laden

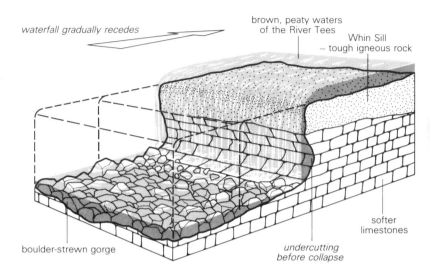

waterfall gradually recedes

brown, peaty waters
of the River Tees

Whin Sill
– tough igneous rock

softer
limestones

undercutting
before collapse

boulder-strewn gorge

Engraved for The Complete English Traveller.

View of the remarkable CATARACT, on the River Teese, which divides the Counties of York & Durham.

course, but in winter it may rapidly turn into a foamy torrent, and occasionally the entire force may freeze, as in 1929 and 1947.

At High Force, the river Tees falls in a single sheet some 20 m into the swirling pool below. Draining from the peat-covered Pennines to the north, the water is dark brown in colour owing to the high content of acids drawn from the peat. But tumbling over the fall it erupts into white foamy plumes, stark against the dark bands of rock framing the spectacle; the foaming waters have been likened to brown ale! Two types of rock can be seen at the fall. The lower has horizontal layers and it forms part of the Carboniferous Limestone. The upper set of dark-coloured rocks has vertical joints within it and is much denser and harder than the underlying limestones. In fact, it is another outcrop of the Whin Sill (see Site 1), composed of the same igneous rock (dolerite). The tough band of dolerite is very resistant to erosion, and this explains the existence of the waterfall, for the river cannot erode this as easily as the limestone. Where there is no dolerite, the valley broadens out and waterfalls are absent, but other bands of dolerite give rise to waterfalls a short distance up stream at Cauldron Snout (NY 814286) and down stream at Low Force. These hard bands of resistant rock, therefore, prevent the river from following a smooth sloping profile through the countryside from source to mouth.

The narrow valley and pebble-strewn valley floor down stream from the fall also owe their existence to the layer of igneous rock. As the river plunges over the waterfall, the softer Carboniferous rocks at the base of the fall are, in time, hollowed out from beneath the hard dolerite layer. The well jointed dolerite, now undermined, proceeds to collapse, filling the valley with debris which is smoothed and rounded into boulders by the action of the water flowing over its surface. The progressive

The fells between Alston and High Force

collapse of the dolerite leads to the retreat of the waterfall up stream, leaving the narrow gorge-like valley cut through the dolerite. This valley shape is very different from that 1.5 km down stream where the dolerite band is absent.

Entrance to the High Force is opposite the High Force Hotel (NY 885286) on the B6277 between Alston and Barnard Castle. The beautiful drive from Alston to High Force across the fells is highly recommended. Small deserted mines and settlements, and peat deposits, can be seen along the way.

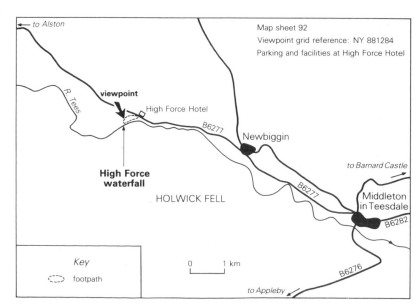

4   Helvellyn, Striding Edge and Red Tarn

If the ice advanced across Britain today, it would demolish all traces of our civilization as easily as we sweep the garden path. Later on, as the ice melted, it would regurgitate a sad and unrecognisable jumble of fragments from everyday life. Today, in England and Wales there are no glaciers, but their power during the Ice Age is revealed by the dramatic valleys gouged from the volcanic uplands of the Lake District.

Excellent examples of the scouring effects of glaciers can be seen around the hollowed-out slopes of Helvellyn, one of the higher peaks in the area.

Striding Edge (foreground) and Red Tarn nestling in the Helvellyn corrie

Starting out from Patterdale, which nestles near the shores of Ullswater, scramble up Helvellyn by way of Striding Edge, and then return along Swirral Edge to the path that leads back to Patterdale. Following this route you will walk along narrow ridges (Striding and Swirral Edges) in places less than 3 m wide, with massive slopes of boulders (called 'scree' slopes) falling away steeply on either side. The ridges outline the top of a giant armchair shape (a corrie) carved into the mountain side. In place of the giant's seat-cushion there is a small lake, Red Tarn, that lies in the base of the huge hollow. The summit of Helvellyn, at 950 m, is marked by a triangulation point, and is the approximate half-way mark of this energetic hike.

In the Ice Age, glaciers accumulated in the higher parts of the Lake District. On the cold north- and east-facing slopes sheltered from the Sun, the glaciers developed strongly, and this explains the fact that corries occur on the eastern side of the Helvellyn mass. By contrast, the slope westwards down to Thirlmere is much more gentle and is without the deep gouges and little tarns witnessed on the journey up from Patterdale. This tendency for ice to accumulate on cold easterly slopes was strengthened by the snow-laden winds coming in from the west. As the wind passed over the mountain top to the easterly side, it developed a backwards swirl, dumping snow on the cold upper part of the sheltered east-facing slope. The presence of the backwards swirl can often be detected on the waters of the tarn, where wavelets move towards the backwall of the corrie basin.

The glaciers ground and quarried the rock and produced deep hollows. Some of these corries subsequently became the sites of such lakes as Red Tarn and Grisedale Tarn. Red Tarn

The shaping of Helvellyn and Red Tarn

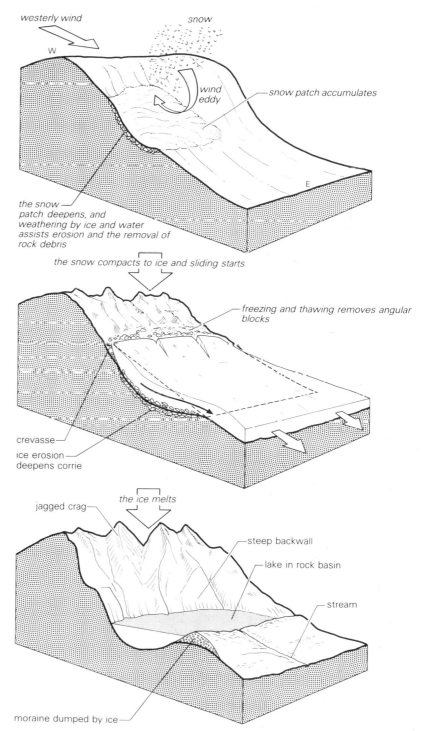

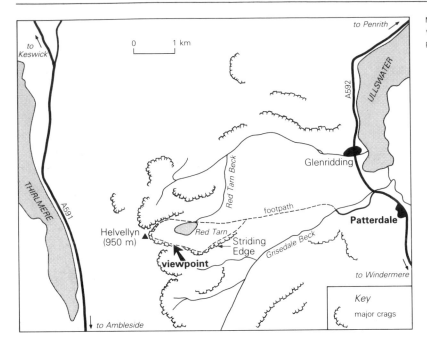

Map sheet 90
Viewpoint grid reference: NY 345150
Parking and facilities in Patterdale

itself is 26 m deep and 500 m across, and in 1860 a low dam was built of boulders, raising the water level by some 3 m in order to supply power to the lead mines in Glenridding. When two or more such hollows occur close to each other on the flanks of a mountain, their enlargement during glaciation reduces the body of the mountain. Sometimes this reduction is so extreme that only a narrow, precipitous ridge remains between the corries. Such ridges are called 'arêtes'; Striding Edge and Swirral Edge are excellent examples.

The splendour of the area has inspired many thousands of visitors. If the complete hike is too demanding, take the easy walk up to Red Tarn, from which you can view the distant parties of hikers clambering along Striding Edge, appearing much like a column of ants. If you are planning to walk up to Helvellyn, then do check on the weather forecast, since mists can descend very rapidly. It is also advisable to wear stout footwear and to take the Ordnance Survey map (sheet 90), a waterproof jacket and a little sustenance.

On a good day the walk is exhilarating and the scenery spectacular. One of the more famous visitors, William Wordsworth, felt moved to express it thus:

It was a cove, a huge recess,
That keeps, till June, December's
  snow;
A lofty precipice in front,
A silent tarn below!
Far in the bosom of Helvellyn,
Remote from public road or
  dwelling,
Pathway, or cultivated land;
From trace of human foot or hand.
There sometimes doth a leaping fish
Send through the tarn a lonely
  cheer;
The crags repeat the raven's croak.
In symphony austere;
Thither the rainbow comes – the
  cloud –
And mists that spread the flying
  shroud;
And sunbeams; and the sounding
  blast,
That, if it could, would hurry past;
But that enormous barrier holds it
  fast.

(from *Fidelity*)

5   Wastwater Screes: a Lakeland backdrop

On a clear summer's day Wastwater shimmers, its deep-blue colour reflecting the sky. Boulders piled one on another up to 300 m above the lake provide a steely grey backdrop. Beyond the boulders and high above the lake extend the moorlands, and to the north-east, Scafell, one of England's highest mountains at 975 m. Occasionally the silence of the lake is shattered by the sharp crack of a falling boulder echoing across the valley; but otherwise it is tranquil.

Almost all of the steep, eastern valleyside is covered by the coarse debris of boulders called 'scree'. It even extends beneath the lake in places, but it never quite reaches the top of the valleyside. Mostly, 100 m or so of rock is exposed as a cliff above the scree. This cliff (the source of the scree) is termed the 'free face'. Thus, the cliff and the boulders are composed of the same rock, which in this case is volcanic material dating back to the Ordovician period.

The boulders can be detached from the cliff in several ways. Probably the most important of these is the action of frost. Water freezing in cracks in the rock will expand and force it apart, eventually prising off blocks. This is a process of physical weathering (breakdown), and its importance is revealed by the angular shape of most of the blocks. Chemical breakdown may also detach blocks, especially where water penetrates along joints in the rock and dissolves or weakens mineral grains. Chemically weathered blocks often show signs of rounding. In either case, the size of the blocks detached will depend on the spacing of the joints and planes of weakness in the rocks. Some boulders are over 2 m in diameter, although the majority are much smaller. In general, the larger boulders are found at the base of the scree, because their weight carries them further than smaller blocks. All the boulders, once detached, settle at a stable angle under the force of gravity. The larger the blocks, the steeper the angle, and at Wastwater it is very steep (35°). Settling results in the scree becoming well packed.

A careful look at the valleyside will reveal that the blanket of scree is made up of a number of separate fans, each spreading downwards and outwards from a trench or gully in the cliff face. Where there are weaknesses in the cliff face, such as joints or faults, the boulders are more easily detached, and so more are removed, leaving a gully in the cliff

face, with the boulders falling as an apron of scree beneath the gully. It is common for adjoining fans to mix together and overlap towards the base, forming an almost complete blanket of scree over the side of the valley. Once covered by scree, the rocks in the valley are protected from further attack, whereas the uncovered upper slope is gradually worn backwards as boulders fall away from it. Therefore, as the scree builds upwards over time, an unusual convex shape of valleyside is formed and preserved beneath the scree.

The screes have formed since the valley was scoured and deepened by glaciers during the Ice Age. Deepening occurred to 15 m below present sea level, forming a rock basin that has since been flooded, giving a lake. The lake is also dammed by glacial clays (moraine) deposited at its lower end, near Wasdale Hall. The scree, like the lake, has therefore been formed fairly recently. The constant addition of boulders, the uneven steep surface and the relatively harsh climate

prevent the formation of soil, and so the scree has not been colonised by vegetation.

The screes are best observed across the lake, from its southwestern edge. For those interested in a closer inspection, there is a footpath along the base of the scree slope. The lake itself may be approached easily from the west, but by far the most scenic route is from Ambleside across the Wrynose and Hard Knott passes. However, this road is single track in many places (with passing points), and there are gradients of 1 in 3 and 1 in 4 near the passes. In the 1820s the route was thought to be so precarious that the traveller, Edward Baines, explained why he and a companion left their ladies behind:

'This step may appear so outrageously ungallant as to require explanation. The excursion to Wastwater, then, be it known, comprehends some of the most

The nature and development of scree slopes

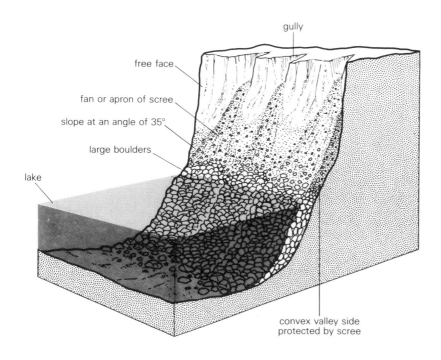

gully

free face

fan or apron of scree

slope at an angle of 35°

large boulders

lake

convex valley side protected by scree

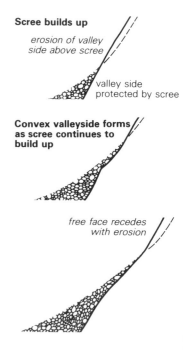

Scree builds up

erosion of valley side above scree

valley side protected by scree

Convex valleyside forms as scree continues to build up

free face recedes with erosion

sublime scenery in Cumberland ...
Very few persons make the
excursion, as it is long, and leads over
rugged tracks; but this made us more
desirous of accomplishing it. Not
having seen that part of the Lakes
before, I was anxious to add it to my
stock of knowledge respecting them;
and George was ready for any
project, especially any which would
take us into the wilder parts of the
country. Nor was he at all sorry, as
he uncivilly expressed it, to get rid of
the encumbrance of his aunt and
sister; for if we went into Wastdale,
it was next to impossible that they
should accompany us, owing to the
length and difficulties of the road.'

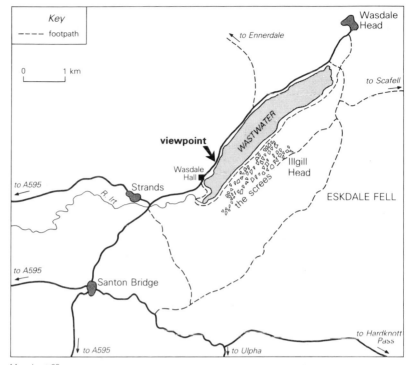

Map sheet 89

Viewpoint grid reference. NY 140048

Limited parking on road side

6 Diamonds in limestone at Hutton Roof Crags

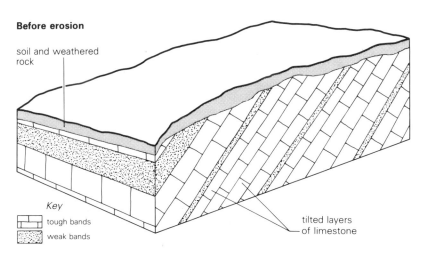

Before erosion

soil and weathered rock

Key

tough bands

weak bands

tilted layers of limestone

ice erosion scours and exposes the limestone surface

**After erosion –
the stepped pavement**

pavement with diamonds (see photograph)

easy erosion of weak bands creates stepped pavement

Formation of the limestone pavement at Hutton Roof by glacial erosion of dipping rock layers followed by solutional attack

Hutton Roof, a small and peaceful village in Lancashire, hosts one of the most striking examples of a limestone pavement in Britain. Follow the pathway from beside the church, through the patch of woodland and out on to the unexpected barren wastes of the pavement. Useless agriculturally, since there is no soil on the precipitous exposed rock pavement, its real value is scenic, but as yet it is little visited. By far the most striking parts are those nearest the church, although the pavement extends over an area of 2.5 square kilometres, and there are other pavements nearby at Newbiggin Crags (SD 545795).

The agile and energetic can scramble easily over the smoothed rock surfaces. For others there is a small grassy pathway: after coming out from the woodland, turn left down the hillside and walk around the end of the exposed rocks. Continue along the base of the rocks for a short distance (50 m or so) until you reach a large boulder (an 'erratic') perched on top of the rock pavement, and then turn right and walk up the hillside, through the spectacular limestone layers. To the left and right, steeply sloping rock faces can be seen, smooth and unvegetated except for plate-shaped patches of lichens and heathers.

The rock is Carboniferous Limestone, massive and relatively hard, but well jointed. The bands of rock were uplifted from the sea bed and folded by earth movements over 250 million years ago, so that now they tilt at 25°, about the same angle as the hillside. After uplift, soil and debris gradually formed at the surface, and then something happened to expose the pavement,

26

removing the soil and possibly some rock, and creating its stepped character. It is thought that the large-scale erosion needed to expose the harder rock surfaces has resulted from the action of ice. Stand at the base of the pavement and imagine an ice sheet, maybe several hundred metres thick, moving over the top of the hill and down towards you. On the way it would tear at and pluck off the soil and debris, revealing the smooth surface of the rock. You may see that the pavement is covered in a few places by a brown sand & clay deposit containing small round pebbles. This boulder clay was deposited by the ice sheet as it melted, but much of it has been washed away since by rain.

The regular slope of the pavement is caused by erosion along planes between the tilted layers of rock. This is helped if weaker bands of limestone separate the stronger bands. Erosion along several weaker bands may give rise to a stepped pavement, as the stronger bands are slowly undermined and will eventually collapse.

You will see, on closer examination, that the surface of the pavement is not smooth in detail but is broken up by deep and narrow ravines (0.3 m in diameter and up to 2 m deep) called grikes, and by smaller (about 10 cm in diameter) rounded channels called rills. These have mostly formed as a result of chemical solution of the limestone by

water, since the glacial excavation of the pavement. Acidic water, especially flowing off peat, is very good at dissolving limestone. Solution may occur on an exposed surface, although it can be more effective under a thin cover of soil or peat, which acts as a sponge. The absence of peat on the surface today may have been caused by climatic fluctuations or by man's activities.

The narrow grikes are formed by solution and erosion along weaker areas, usually joints, where the water can penetrate easily. Once formed, they act as drainage channels and are soon deepened. The grikes at Hutton Roof are particularly fine because they exploit two sets of joint patterns in the limestone, leaving diamond-

Part of the Hutton Roof limestone pavement, showing its well developed grikes, clints and rills

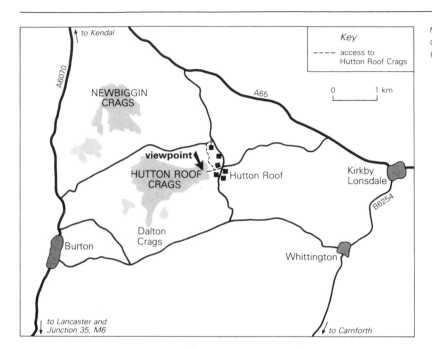

Map sheet 97
Grid reference viewpoint: SD 567782
Parking in village

shaped areas of rock between them. The areas remaining between the grikes are called clints, and their size depends directly on the spacing of the grikes and thus the jointing of the limestone. The rills are also formed by rainwater solution, but their pattern and density are generally controlled by the spacing of the grikes and the slope of the pavement. On steeply sloping pavements, such as at Hutton Roof, the rills are roughly parallel and they drain more or less directly down slope into grikes, whereas on flatter pavements they form more irregular networks and may end in miniature rounded hollows ('swallow holes'). Both rills and grikes drain the surface of the pavement, so where the grikes are widely spaced, rills are often more dense, to cope with the drainage of the blocks.

Solution will eventually destroy the pavement's surface, a bouldery limestone debris and soil being formed as more and more of the surface is eaten away. The full cycle of pavement formation would then be completed by erosion and removal of the rubble during a glacial phase, exposing a new pavement at a lower level.

7   Gaping Gill and the Norber Erratics

The attractive little village of Clapham (SD 745692) nestles on the edge of the Craven lowland and the Yorkshire Dales National park. It straddles a small river, the Clapham Beck, which drains Clapdale. Walking up the dale, following the signed route from the northern end of the village, one passes through woodland with flamboyant rhododendrons, exotic bamboos and Victorian follies, before reaching Ingleborough Lake – a man-made creation. The walk thus far has an 'olde worlde' charm, but this changes rapidly at the head of the lake, where the path climbs steeply towards the limestone plateau, crossing the scarp of the North Craven Fault. Over the next kilometre or so various caves, formed by the solution of the limestone, are found near the path. These include Cat Holes, Clapdale Grotto and Ingleborough Cave (SD 754710), a show cave whose entrance opens at the edge of a 20 m cliff on the west flank of the valley. Just beyond its entrance is Clapham Beck Head, where the beck emerges from its underground passage. Above this point the valley is dry and it is joined by a tributary dry valley, Trow Gill. This is a narrow limestone gorge carved along the weak line of a major joint by surface water flowing during the Ice Age. Under the tundra conditions that existed then, as in Siberia today, the present-day underground drainage systems (some 100 m below) were choked by ice and glacial debris, and so the water flowed on the surface. The path continues along Trow Gill, towards the head of the valley, where there are some impressive pot holes such as Bar Pot, Flood Entrance Pot and Disappointment Pot, but they are nothing in comparison with Gaping Gill, a little further on.

The eastern slopes of Ingleborough, composed of

A giant Norber erratic on its pedestal

impermeable shale coated with boulder clay, are drained by several tributaries which merge into a sizeable stream, Fell Beck. When the stream flows across the boundary between the shale and permeable Carboniferous Limestone, it starts to lose water through the innumerable cracks and joints in the limestone before disappearing abruptly down a remarkably large oval hole – Gaping Gill.

This is probably the best known pothole in England. The entrance takes the form of an oval shaft 20 m by 10 m that lies at the bottom of a box-shaped blind valley 10 m deep, cut in boulder clay and underlain by limestone. Down this shaft, when in flood, the Fell Beck falls 110 m to the enormous main chamber, widening its pothole in the process. The main chamber is the largest cave in

Britain, some 150 m long, 30 m wide and rising over 30 m to the point of entry of the main shaft. Over 11 km of underground passages connect up with the main chamber, and it is along some of these that the water of the Fell Beck travels, before reappearing at Clapham Beck Head.

The first known efforts to climb down and explore carefully took place in 1872 when John Birkbeck of Settle dug a trench nearly a thousand metres long to divert the waters of Fell Beck which usually pour into the chasm. The first complete descent was made by M. E. A. Martel of Paris on 1 August 1895.

From Gaping Gill, instead of retracing your steps to Trow Gill and Clapham Beck you can head eastwards across the plateau to Long Scar and then southwards to Norber (SD 764698), to the famous limestone pavements and erratics. The pavements – rock surfaces exposed by glacial scouring – are similar to those at Hutton Roof (see site 6) except that they are nearly horizontal, reflecting the horizontal layering of the rocks here.

Unfortunately, many of the finest examples of pavement in Britain have been ripped up by man to beautify garden rockeries, but those in this area still remain, and provide important little niches for many species of plant. As Kendall and

Wroot remarked in their splendidly comprehensive *Geology of Yorkshire* (1924),

'Sometimes a mere knife-edge of rock is all that remains of the block between the fissures. On the surface

A view from above showing the main landscape features of the area depicted on the map

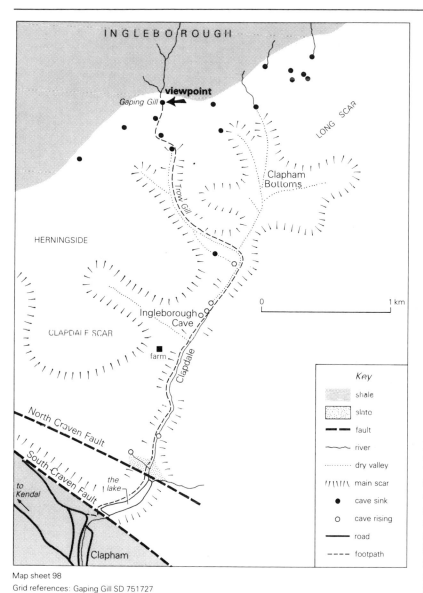

Map sheet 98

Grid references: Gaping Gill SD 751727
              Norber Erratics SD 766697

Facilities: parking and shops in Clapham
          Ingleborough show cave

no living thing appears save here and there a gnarled and wind-haggled hawthorn bush, but in the clefts are luxurious growths of harts-tongue fern and other shade-loving plants – wood-sorrel, wood-garlic, geranium, anemone, rue and enchanter's nightshade.'

Huge boulders resting on pedestals of limestone indicate the amount of solution that has taken place on the limestone surfaces since the retreat of the ice sheets. These boulders are often composed of Yoredale or Millstone Grit and were dumped by the ice as it melted. They are called erratics, because they are made of rocks different to those in the area they are found. Sometimes the erratics have been transported many hundreds of kilometres, but in this case they are fairly local in origin. The erratics themselves are much less soluble than the limestones. In addition, they have protected the underlying pavement from rainfall, rather like an umbrella, so that they are now perched on pedestals of protected rock up to 30 cm high.

These chunks of grit were probably deposited about 12 000 years ago. From the height of the pedestals we can estimate that the limestone surface has been lowered since then at an average rate of about 25 mm each 1000 years. They are one of the finest groups of erratics in Britain, and there are so many boulders that it is difficult to miss them. If you do not feel like walking from Gaping Gill, they are reached easily by the footpath off the lane leading north from Austwick village.

8   Malham and the Pennines

Some of the highest summits in the backbone of England – the Pennines – occur in the Craven Uplands. The peaks of Whernside, Ingleborough and Penyghent, built of terraced layers of hard gritstone and softer shales, rear their table-like heads above the surrounding limestone moorlands and the intervening dales. These rocks weather to give poor soils and dark aprons of 'scree'. They are often capped by a crown of dark brown peat, and they provide the gathering grounds for many streams which rise on their barren, windswept upper slopes and then plunge rapidly into great vertical shafts (such as Gaping Gill, site 7) once they cross onto the porous and permeable limestone.

The uplands are sometimes abruptly terminated by awe-inspiring scars or cliffs – escarpments that have been fashioned along great rock fractures called faults. Thus the darkness of the grit moorlands contrasts with the whiteness of the limestone scars, and the uplands contrast for their part with the well tended farmlands of the Craven lowlands.

Every year thousands of students, young and old, come to the Field Centre at Malham to see these contrasting landscapes, with some of the most spectacular limestone scenery in the British Isles. The small lake, Malham Tarn (SD 895665), is a good starting point. This rests on impermeable Silurian siltstones, covers approximately 62 ha and is mostly less than 3 m deep. Water from the tarn flows out through a hummocky mass of deposits (SD 893662) laid down in the Ice Age. More or less along the line of the road the rocks at the surface change because of the presence of a major fault – the North Craven Fault. The tarn water sinks undramatically into the Great Scar Limestone (SD 895655) at Water Sinks. Most of it

The grandeur of Malham Cove

today reappears at the Airehead Springs (SD 902622), some 2 km south of Malham village, but plainly in the past it flowed along the striking valley that leads southwards – the Watlowes Dry Valley (SD 895647). This valley has stepped sides caused by the varying resistance of the different limestone beds and it may have been carved in the Ice Age when glacial melt waters were present and when the underground joints and chambers in the limestone were blocked by ice frozen in the ground (permafrost). There is evidence of glacial action on the high ground to the east of the valley where there are many areas of exposed limestone pavement created by the planing action of the glaciers. The valley is variable in width and it slopes unevenly, providing something of a scramble in places over angular boulders. Before it turns left there is the site of an old waterfall, around which a small detour has to be made.

At the far end of the valley you will find yourself standing on top of a remarkable great curved grey cliff – Malham Cove (SD 899642) – created when a former waterfall eroded the rock backwards from the Mid-Craven Fault. It is over 100 m high and, on rare occasions after exceptional rainfalls, water overflowing from Malham Tarn reaches the top of the Cove and surges over in a vast sheet. Normally, however, water emerges from a broad, shallow cave at the base of the Cove, but contrary to what one might expect the water does not for the most part come from Malham Tarn but from the Smelt Sink (SD 882659). This is because underground connections in the limestone are complex and are related more to fissures and joints than to the surface relief.

A rough path leads from the top down to the base of the Cove. For the less energetic, parking is available in Malham, and it is but a short walk to the base of the Cove. Along the way, just down valley from the Cove, you will see the old cultivation

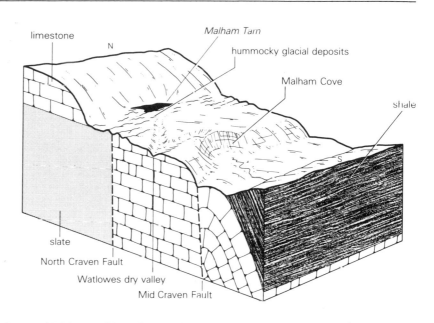

A simplified diagram of the Malham area showing the major landscape and geological features

terraces, called lynchets, that lie on the eastern side of the beck. They were made in the 7th and 8th centuries by Anglian farmers to facilitate ploughing with oxen. There are also some remains of field boundaries from the Iron Age.

The whole setting of Malham Cove impressed many early visitors. John Hutton, for instance, in '*A tour to the caves*' (1780) recorded his sensations thus, though the comparison he draws with Niagara is something of an exaggeration:

'The rocks lie stratum upon stratum and on some there are *Sasca sedilia* or shelves, so that a person of great spirit and agility but of small and slender body, might almost walk round. A small brook springs out of the bottom of the rocks; but in floods the narrow subterranean passage is not able to give vent to all the water, when there pours down a stupendous cataract, in height almost double that of Niagara.'

On the other side of Malham village there is another deep cleft that is well worth a visit: Goredale Scar, described by Wordsworth as 'Goredale chasm, terrific as the lair where the young lions couch'. This can be approached along Goredale Beck by a good footpath. The scar differs from Malham Cove in that water still flows over the top, plunging down in two impressive waterfalls. The waterfalls are of particular interest because they are covered with great masses of a pale-coloured substance called tufa. This forms because large quantities of limestone are dissolved in the waters of Goredale Beck. As the water breaks over the waterfalls, carbon dioxide gas is given off, causing the dissolved material to be redeposited as tufa, another form of limestone. The top waterfall gushes through a hole in the rock, a hole originally blocked by debris; and in the valley above the top fall there used to be a deep and narrow lake. In 1730 a violent flood broke through the debris, causing the lake to drain and the stream to be diverted to its present course. The original route of the stream flowing from the lake is marked by an old tufa screen to the left of the hole. This fascinating material is very fragile, and will be irreparably damaged if visitors either collect samples or clamber across it.

Like Malham Cove, Goredale Scar impressed early visitors who found it horrible, formidable and uncomfortable. A Mr Gray penned his feelings in October 1769.

'As I advanced, the crags seemed to close in, but discovered a narrow entrance turning to the left between them; I followed my guide a few paces, and the hills opening again into no large space; and then all further way is barred by a stream that at the height of about fifty feet, rushes from a hole in the rock, and spreading in large sheets over its broken front, dashes from steep to steep, and then rattles away in a torrent down the valley; the rock on the left rises perpendicular, with stubbed yew trees and shrubs starting from its sides, to the height of at least 300 feet; but these are not the thing; it is the rock to the right under which you stand to see the fall, that forms the principal horror of the place … The gloomy uncomfortable day well suited the savage aspect of the place, and made it still more formidable; I stayed there, not without shuddering, a quarter of an hour, and thought my trouble richly paid for; for the impression will last for life.'

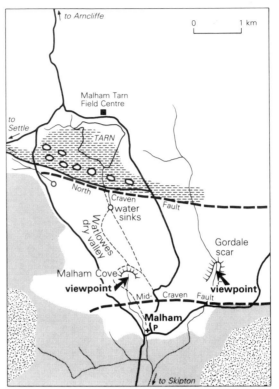

Key

— fault

sandstone

shale

slate

hummocky glacial deposits

P   car park

---- footpath

9 Newtondale: an Ice Age torrent

Walking or driving across the wild and windswept Cleveland Hills in the North Yorkshire Moors, you may be surprised to encounter a deep sinuous trench, as much as 250 m wide and 100 m deep, with very steep sides. Newtondale, as it is called, stretches from Pickering northwards for 15 km towards Whitby. On further inspection you may ask, why does such a large winding valley contain such a small stream? Indeed, at its upper end there are no streams whatever, only a great peat bog and some artificial drains. It is impossible for such a small volume of water to have carved this imposing feature, although its meandering course, like that of many rivers today, suggests that it was formed by running water – but how and when?

We have to go back to the Ice Age to answer these questions, to a time when great floods of water were sufficient to erode the giant channel. An early theory for the origin of the floods came from a geologist, Percy Kendall, at the turn of the century. He imagined the advance of a large glacier or ice sheet, from the east and the north, until it abutted onto the higher hill mass of the Cleveland Hills and almost encircled them. This would prevent the rivers from draining to the sea, and they would be impounded by the ice to form a series of lakes in Eskdale, Glaisdale and Wheeldale. These glacier-dammed lakes would become deeper and deeper until their level reached that of the surrounding countryside at about 200 m above present sea level. Then the lakes would overflow in great torrents of water, carving a giant channel to the south along a line of pre-existing valleys. Thus, the water and its sediment load would be carried towards Pickering, where the lowland (Vale of Pickering) may itself have contained a huge lake dammed by ice (Lake Pickering). Kendall argued that when the water reached the Vale of Pickering it dumped its load in the still lake to form a great delta covering over 5 square kilometres.

More recently, Kendall's evidence to support his theory, such as old lake shorelines, lake sediments, channels and the huge delta, has been questioned, and a simpler idea has been put forward to explain Newton Dale. The suggestion is that, as the climate warmed at the end of the glacial period, the ice sheet decayed, releasing large amounts of water, which carved the channel. This huge meltwater channel, so impressive to look at, would have been carved in

The Newtondale channel, cut by raging Ice Age torrents

one, or at most a few, decades by massive volumes of water – possibly a flow of about 10000 cubic metres per second, ten times the present flood discharge of the Thames.

Not very far from the great channel at Newtondale, on the south edge of the moor about 4 km ENE of Levisham, there is a splendid series of isolated rocks (or tors) sculpted from Jurassic grit deposits. Like sentries, the tors overlook the valleys of Bridestones Griff and Dovedale Griff (the latter is not to be confused with Dovedale Gorge, site 13). There are two main groups of tors, the High (SE 873914) and the Low (SE 874912) Bridestones, respectively north and south of the Bridestones Griff. An immense variety of fascinating shapes are to be seen, including the well known 'mushroom' or 'pedestal rock' form. Now owned by the National Trust, the Bridestones can be approached along several tracks leading off the A169 near Lockton. The most attractive route follows the 'Forest Drive' and the reader is referred to OS Sheet 94 for further details. There is ample parking.

Undoubtedly the tors owe their existence to the resistant Jurassic grit, which is heavily jointed. This rock, found near the top of the hillslopes,

overlies less resistant material. The local streams have attacked the underlying weaker rocks, thus undermining the grits, and they have also eroded down and back along joints in the grit to expose the wierdly shaped tors. Wind sculpting and frost weathering in the Ice Age may have helped to create the shapes.

Broadly comparable tors occur within another National Trust site at Brimham Moor (SE 210647) about 12 km SW of Ripon. But here it is possible that man may have helped to shape some of the more bizarre forms, for in the 18th century any little defect or inelegance in nature was readily remedied by the

landscape gardener. Whether it was the rain working on beds of unequal resistance, fierce winds blasting away at the weaker beds, or the hand of romantics in the 18th century – the rocks, scattered over some 20 ha of Brimham Moor, present a great variety of weird and wonderful shapes. Tourist guides have been wont to point out gigantic mimic heads of elephant, hippopotamus and porpoise, as well as dancing bears, 'Druids' sacrificial basins', 'Druids' reading desks', oyster shells, and as the visitors become more and more intimate they are introduced to 'kissing chairs' and 'lovers' leaps'.

Key

land over 200 metres

steep sides of Newtondale

National Park boundary

Map sheet 94

Viewpoint grid references:
SE 822913 or SE 855982

Earlier this century, Percy Kendall proposed that Newtondale was the spillway or overflow channel of water impounded in lakes by a great ice sheet

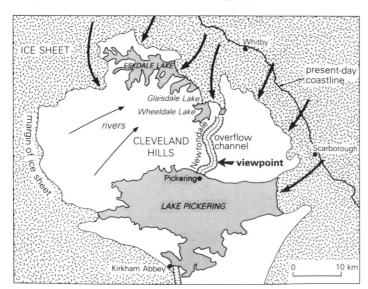

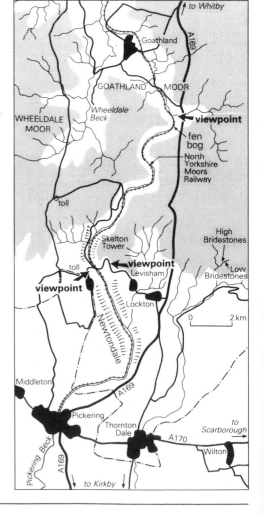

10 Hook across the Humber at Spurn Head

Spurn Head pokes its curving finger of sand and shingle a good third of the way across the mouth of one of Britain's largest and muddiest estuaries – the Humber. Its 'roots' are near Kilnsea in a wedge of glacial and river debris deposited in the Ice Age. From here the sand spit sweeps round 5 km on a southwesterly course, its thin neck widening into a pear-shaped end. The seaward edge of the spit is straight, whereas the estuarine side is more irregular.

The material making up Spurn Head is derived from the fearful erosion of the cliffs of Holderness that lie to the north. Strong winds off the North Sea attack the soft glacial boulder clays that compose the cliffs. Observations show that over the past century the rate of retreat has been as much as 2.75 m each year; this is among the highest reported rates of cliff retreat in the world, over such a long period. Many villages, reported in *Domesday* and other historical

documents, have disappeared, and probably something over 200 square kilometres (equivalent to a strip of land about 4 km wide) have been lost since Roman times.

A portion of this eroded material travels southwards along the coast under the influence of waves and currents (a process called 'longshore drift') and accumulates as Spurn Head. However, the process of accumulation has not been continuous, for the spit also suffers

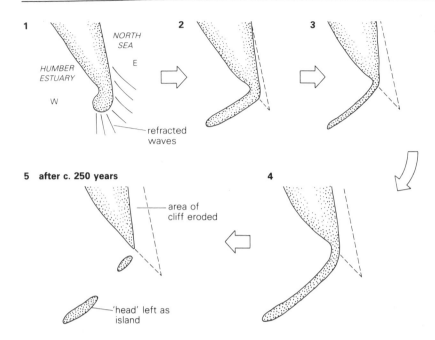

Stages in the growth and decline of Spurn Head

periodically from the ravages of storms. Historical information suggests that episodes of destruction at intervals of around 250 years have punctuated considerable periods of persistence and growth of the spit. Major breaches occurred at about 1360, 1610–20, and in the years 1849–56. Thus, the growth and destruction of the spit appears to go in regular cycles.

During the early stage of a cycle, the youthful spit gradually grows southwestwards into the sheltered area protected by its root. The angle between spit and coast is fairly sharp, and the spit is curved round by waves refracted in the shallow sea. As the cycle advances, the spit continues to grow, but both the coast and spit are eroded on the seaward edge and they retreat westwards under this attack. The root is thus shortened and the spit lengthens both northwestwards and southwestwards, leaving the thin neck more and more exposed to

vigorous wave attack. Moreover, as the spit grows it is not reinforced by material travelling up the inside edge, for the smaller waves in the estuary can only carry a short distance the material that rounds the tip of Spurn Head. The seaward side continues to retreat, and that section of the neck which is deprived of support from the rear becomes progressively thinner and very vulnerable to major storms. Once a breach occurs, it is scoured out by the flood tide. The Head is left as an island which, being deprived of its source of fresh material from Holderness, soon disappears. The cycle starts again when a new spit grows out, but farther to the west than its predecessor, since the former coastline has been eroded away.

Spurn Head can be approached along a track from Kilnsea (TA 410158), although its full splendour is best appreciated from the air.

11 Alport Castles

The survival of the unspoilt Peak District, following close on the heels of urban and industrial chaos in Sheffield, always comes as a very pleasant surprise. In fact, the drive from Sheffield to Glossop along the A57 is simply delightful, passing by the man-made expanse of the Ladybower Reservoir and close to the high Peak and through Snake Pass. However, to appreciate this countryside fully you need to take to the footpaths and tracks, and none is better than that along the beautiful Alport valley. The main footpath begins beside the A57 just west of the Alport River, near Hayridge Farm (SK 141896).

As you walk up the valley, many interesting features can be seen. Active erosion by the small meandering stream has led to the formation of river terraces and river cliffs. The valley is eroded through weak shales, which are overlain here by tough Millstone Grit, similar to that observed at Brimham (site 9) and Ramshaw (site 14). The hard, but well jointed grits are visible along the far (eastern) side of the valley, forming an imposing cliff (or 'scarp') on the skyline. The grit gives way to shale down the valley side, the boundary sometimes clearly shown by small springs seeping out between the permeable grit and the impermeable shale below.

This eastern valley side and the grit cliff repay closer investigation. Follow the footpath as it crosses Alport River (near Alport Castles Farm) and climbs up and across a huge mass of hummocky material to the cliff behind. It becomes obvious when you reach the top of the cliff that great blocks of grit have broken off from the main scarp face and have slid down the side of the valley. The blocks are thus separated from the steep cliff face by narrow chasms

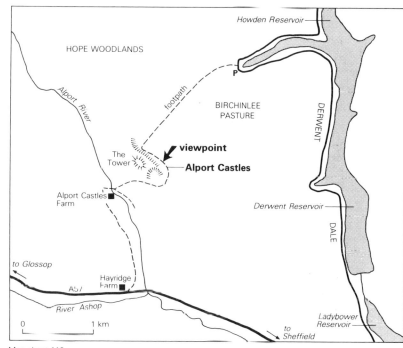

Map sheet 110

Viewpoint-grid reference: SK 144914

Parking (**P**) grid reference: SK 154927

Roadside parking near Hayridge Farm

littered with angular fragments of grit. It is from their resemblance to castles, possibly 'protecting' the valley, that the blocks have derived their name. Once you have reached the top of the cliff, a short walk along it in a northwesterly direction will reveal spectacular views of the rock castles, and of the hummocky ground over which you climbed. It has been suggested that this is the largest landslide in Britain, the sliding taking place at many points along the side of the valley, and resulting in not only the rock castles but also the great lobes of hummocky debris on the side of the valley. The blocks of rock have not been tilted

backwards in the process of sliding, unlike at Mam Tor (site 12). It is not clear whether sliding is still taking place today.

The cause of the sliding is an interesting problem, in which several factors are important. The major structural control is the large-scale jointing in the grits, which enables the blocks to break away and also accounts for the steep cliff left behind. Secondly, the seepage of water, from between the grits and underlying shales which act as supports, may weaken the shales and thus promote sliding. Thirdly, there are processes that occurred during the Ice Age which could have

The tower of Alport Castles separated from the gritstone cliff and rising above the Alport Valley

affected the stability of the shales and hence the grits. It is possible that glacial erosion caused oversteepening of the side of the valley, particularly in the weak shale, with the result that the shale and overlying grit slid down into the valley once the glaciers melted. Also, it has been suggested that, on melting, permafrost in the surface layers of the shale, just beneath the grits, would have caused mudflows which would have carried the blocks of grit with them. Thus, all the evidence seems to suggest that, with one trigger or another, this arrangement of weak, impermeable shale beneath hard permeable grit in the valley side is responsible for the landslides. The structure of the grit controls the size and shape of the castles and enables whole blocks of rock to be removed.

The castles can be approached from Derwent Dale too, following the clearly marked footpath from the western side of Howden Reservoir (SK 154927). The path climbs up and across the grit moorland (Birchinlee Pasture), carpeted with peat and inhabited by sheep. Indeed if you have the time and available transport, and the weather is good, the walk from Hayridge Farm to Howden Reservoir via the castles is exhilarating.

12   Mam Tor: the 'shivering mountain'

Mam Tor, the so-called 'shivering mountain', frowns down on the village of Castleton. Like Alport Castles (site 11), which lie further to the north, it consists of a grit escarpment, of which the upper parts are being undermined by the removal of the softer shales below. The grits are constantly falling and slipping, creating the shivering effect that made Mam Tor one of the original 'wonders of the Peak', and gave rise to its name. According to Daniel Defoe, the 18th century traveller, Mam Tor 'in mountain jargon signifies, the Mother Rock, upon a suggestion that the soft crumbling earth, which falls from the summit of the one, breeds or begets several young mountains below'. Not only is the landsliding particularly active, but it is also highly detrimental to the main A625 road which snakes across the slide. From time to time great cracks develop in the road's asphalt surface, closing the road to through traffic. Indeed it has been closed for some years now, and until it is mended, you will be able to estimate the amount of slippage that has taken place since the photograph was taken.

The landslide covers an area of 35 ha (almost 1 km long and 550 m wide) and is backed by a main scar 105 m high. It is one of the best examples of a landslip to be found in England, and exhibits most of the classical textbook features. The great cliff marks the site of the original rupture, while the uneven ground down slope marks the zone where blocks of rocks have slipped and rotated backwards. Yet further down slope, the disturbance created by the

Map sheet 110
Grid reference of slip: SK 132836

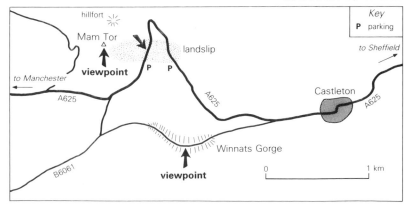

41

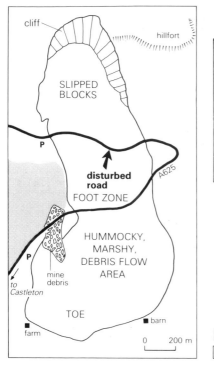

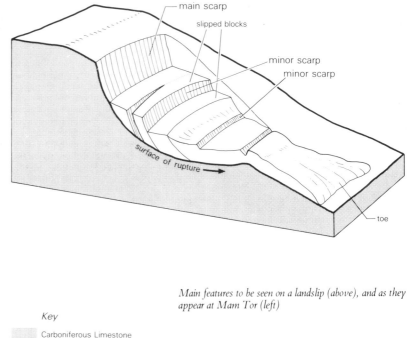

Main features to be seen on a landslip (above), and as they appear at Mam Tor (left)

Key

Carboniferous Limestone

Carboniferous Millstone Grit

slipping blocks has caused debris to flow and spread out as the marshy toe of the landslip.

It is worth climbing to the top of the cliff, for on the top there is not only a large Iron Age camp, but also a remarkable view. From the summit (at 517 m) you can look westwards along Rushup Edge, northwestwards across the Vale of Edale to Kinder Scout (637 m), due north to the Peak, north-east to the glinting waters of the Ladybower Reservoir, and eastwards across the remains of Peveril Castle and Castleton. For those who merely wish to view the landslip from near its base, or who wish to clamber across it, there is some roadside parking along the A625 just before it is blocked off.

For the less active there is, nearby, a small limestone gorge of very striking character called Winnats (SK 136826). A narrow steep road passes along its length, and at its summit provides a fine view of the great cliff at Mam Tor. This beautiful, emerald-carpeted gorge keeps its secrets well hidden underground in caves and caverns. Not only do these contain the rare Blue John stones but also some fine stalactites. The entrance to the show cavern is, unfortunately, signalled by the usual tourist paraphernalia.

13   Dovedale: caverns, spires and arches

The Peak District, the last link in the Pennine Chain, has been called 'the knobbly bottom bone in the spine of England'. Rising up above the heartland of industrial England it provides both solitude and recreation. At its centre is a limestone plateau – the White Peak – seldom below 300 m and often rising to above 450 m. One of the most beautiful parts of this unspoiled area is Dovedale, notable for its rock spires and natural arches and for the excellence of its trout. Isaak Walton, his disciple Cotton, and Sir Humphrey Davy, have all celebrated it, not only for the sport it afforded them, but for its natural charms. It is even said that Handel, while walking beside the Dove, was inspired to begin composing the *Messiah*.

On entering this enchanting spot, the sudden change of scenery is most striking. This contrast was well captured in a description written in the time of King William IV, and it is still true today.

'The brown heath, or richly cultivated meadow, is exchanged for rocks abrupt and vast, which rise on each side, their grey sides harmonized by mosses, lichens, and yew trees, and their tops sprinkled with mountain-ash. The hills that inclose this narrow dell are very precipitous, and bear on their sides fragments of rock that, in the distance, look like the remains of ruined castles ... The objects which, at a distance, appeared to have been ruins, are found to be rude pyramids of rock and grand isolated masses, ornamented with ivy, rising in the middle of the vale. The rocks which inclose the dale ... overhang the narrow path that winds through its dark recesses, and, frowning, in craggy grandeur, and shaggy with the dark foliage that grows out of the chinks and clings to the asperities of the rocks, forms a scene unrivalled in romantic effect.'

The Dove, like many of the Peak rivers, begins life as a mountain stream on the impermeable Carboniferous Millstone Grit. At one time the grits and shales covered the whole of the limestone plateau. Gradually, however, these upper beds were stripped by erosion, and the Dove was forced to cut down into the underlying limestone, though still following its initial course. An increase in the volume of water, due to ice melting in the Ice Age, may have aided downcutting. The river

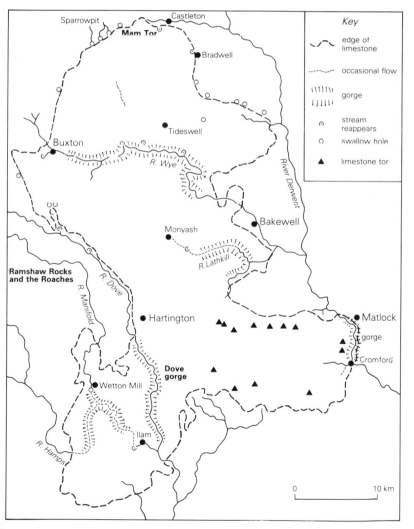

Landscape features in and around the White Peak

Reynard's Cave on the eastern side of Dovedale

The River Dove flows peacefully beneath a pinnacle sculpted from Carboniferous Limestone

cut down deeply, forming a gorge-like valley, with almost vertical cliffs where it encountered hard bands of limestone. A great group of Dovedale features has since been carved out of the limestone exposed in the valley sides and in the spurs that once projected into the gorge. This sculpting was mostly accomplished by frost and by solution of the limestone. The resulting spires and pinnacles, such as the Twelve Apostles, Ilam Rock, Pickering Tor, Lover's Leap and Tissington Spires, are like sentries guarding the gorge.

However, the incision of the gorge and the sculpting of spires and tors are not alone in creating the bizarre landscape features of Dovedale. Since the Dove here flows through limestone, many features result from the dissolving action of the water and the erosion by underground drainage systems so characteristic of this rock type. As the river cut down, it came across old caves and caverns in the limestone, formed at an earlier stage by rain water draining underground through the limestone. These caves and other lines of weakness in the

limestone made the downcutting easier and today they add great variety to the scenery of the gorge. Reynard's Hole and Reynard's Kitchen are the remnants of an old cavern, now exposed and perched high up on the valley side. A natural arch (12 m high and 5.5 m wide) spans their entrance, which is approached along a steep and slippery side-track. This cavern was probably exposed at an early stage in the downcutting. Dove Holes, close to where the tributary Hall Dale joins the gorge, are two more arched recesses in the valley side. The larger has a span of 17 m and a height of 9 m, but it penetrates only a short distance into the hillside.

The walk up the wooded valley of the Dove is easy. The path follows the winding course of the river and, as in the time of King William IV, the visitor is continually presented with new vistas enhancing the dramatic effect of the many extraordinary features. The car park and visitor's centre (open in summer only) near Ilam (SK 146507) provide a convenient starting point. Cross over the river using the bridge near the car park, and walk along the eastern bank: the track along the western bank soon ends and you are then forced to cross the river on stepping stones, which are often under water during times of higher flow.

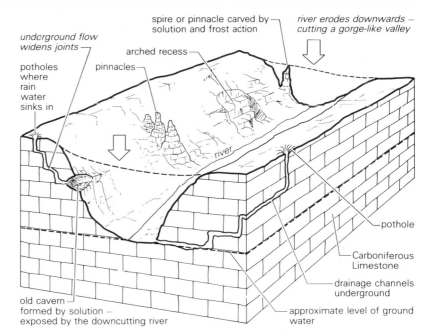

The formation of Dovedale, and its major features

Key

cave	
arch	
spires	
footpath	
high land	

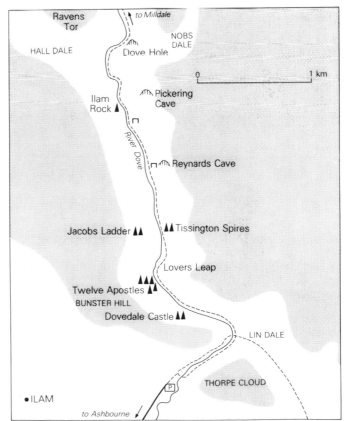

Map sheet 119

Car park grid reference: SK 156509

Facilities: interpretation centre, toilets and refreshments near car park

14   The prodigious gorge at Matlock

One of the most famous landforms of the Matlock area is the feature called Matlock Tor. Daniel Defoe thought it the most wondrous in the Peak:

'On the other, or east side of the Derwent, stands a high rock, which rises from the very bottom of the river (for the water washes the very foot of it, and is therefore in dry weather very shallow); I say, it rises perpendicular as a wall, the precipice bare and smooth like one plain stone, to such a prodigious height ... that which adds most to my wonder in it is, that as the stone stands, it is smooth from the very bottom of the Derwent to the uppermost point, and nothing can be seen to grow upon it. The prodigious height of this tor (for it is called Matlock Tor) was to me more a wonder than any of the rest in the Peak ...'

This is not a tor in the sense of being some isolated stack of rock. Rather it is a great white cliff of limestone, a celebrated feature of the many grand scenic views enjoyed in Matlock and Matlock Bath since the 18th century. You can drive to a cafe above the High Tor from Starkhomes (SK 300589) and see the 120 m sheer fall to the river below, and investigate the narrow gorges and fissures that tunnel through the limestone. Some of these are natural and some the result of the work of lead miners. The wooded setting adds to their charm.

The Matlock Gorge, above which the tor rises, and which was formed by the River Derwent cutting through the eastern edge of the Peak, is in many respects comparable to Dovedale (site 13). It is, however, on an altogether larger and bolder scale, though less sympathetically treated by the hand of man. Even a fleeting visit by car is sufficient to obtain a feeling for this grand chasm, but a

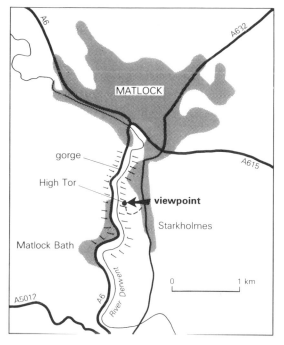

Key
—— roads
— — — track

Map sheet 119
Viewpoint grid reference: SK 297590
Cafe etc. at viewpoint

Gritstone tors at Ramshaw Rocks

more spectacular general view is had from high up on the Derwent Valley side above Matlock Bank near the imposing County Offices (SK 300609).

At Matlock Bridge the river leaves the open valley excavated in the gritstone and shales and, rather than continuing an apparently easier route towards the south-east, plunges into the limestone. Its curious course, as with the Dove, probably results from its superimposition on the limestones from a formerly more extensive outcrop of the less permeable gritstones and shales. As the grits and shales were gradually removed, the river was forced to cut down through the limestone. An alternative explanation is that its course was diverted in the Ice Age by glaciers pushing it from the east.

Elsewhere in the Peak, true tors are common, their formation being favoured by the resistant gritstones. A fine locality to see them is at the Roaches (SK 003630) and Ramshaw Rocks (SK 018620), alongside the Leek to Buxton road (A53). The distinctive character of the area results from the way in which the rocks have been downfolded in the shape of an elongated pie-dish (called a syncline) with two hard grit bands, separated by weaker shales, forming the two distinct rock rims, one within the other. The outer rim, of Roaches Grit, forms the ridge of the Roaches proper and the striking Ramshaw Rocks. The sharp dip of the grit beds means that the knobbly spine of Ramshaw Rocks thrusts itself to the east. The forces of erosion, which have worked on weaknesses in the rock, have carved this spine into many weird shapes. The Roaches appear more castellated and can be reached from the car park at SK 006615 (for Hen Cloud) or from SK 004622.

Map sheet 119
Viewpoint: grid reference: SK 019619
Car parks (P) at SK 00615 and SK 004622

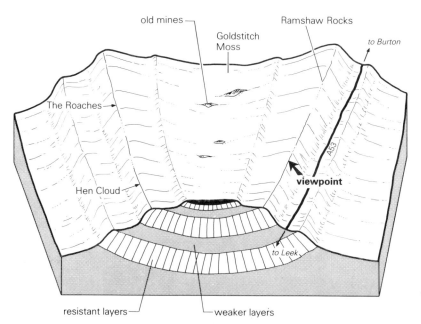

Ramshaw Rocks and The Roaches; simplified diagram to show the pie-dish form of the geological structures

15   Winsford Bottom Flash: the Cheshire Broads

In such a small and crowded country as England, the hand of man in the development of the landscape can be strong and pervasive, particularly where industrial activity has been long and continued. Even in the Middle Ages, in an essentially agricultural area, man succeeded in creating the Norfolk Broads of East Anglia by his peat-digging activities (site 34). In the industrial area of mid-Cheshire, especially from the middle of the 18th century onwards, the exploitation of rock salt – the main form of local mineral wealth – has resulted in some notable man-made landforms. These include the small lakes that are known colloquially as 'the Cheshire Broads'. The most famous of these, and one that has now become a popular venue for sailing, is Winsford Bottom Flash (SJ 657655).

When the Romans invaded England and overran the Midlands, they discovered the natives of Cheshire manufacturing salt by pouring brine upon faggots of charcoal and scraping off the crystals as they formed. More modern methods of production started in the Winsford area in the early 18th century, and the industry was much aided when the River Weaver was rendered navigable in the 1730s. Production reached its peak in the 1880s and in that decade over 7 million tonnes of salt were shipped down the river.

The salt beds are part of the Keuper Series of rocks of Triassic age; the name Keuper is borrowed from similar deposits in Germany. The deposits originated in salty lagoons or desert lake basins during times of hot, dry climatic conditions, over 200 million years ago. The lakes or 'flashes' reflect the subsidence produced by mining the salt. The brine is pumped out of the salt beds, leaving a hollow into which the overlying rocks sink and collapse. The basin formed on the surface as a result of the collapse is soon infilled with water, and a lake is born.

It is well known in coal mining areas that the removal of coal seams from depth causes serious settlement or subsidence of the ground surface. In rock salt areas the extent of subsidence can be even greater for a variety of reasons. First, the rock salt is highly soluble in fresh water, so that, if fresh water entered the old mine workings, it rapidly damaged the already weak supporting pillars of the mines, causing accelerated and widespread collapse. Secondly, this solubility meant that much of the salt was later won by natural brine pumping. That is, the salt, dissolved in the ground water present in the rocks, was pumped out as brine. While the natural brine reservoirs

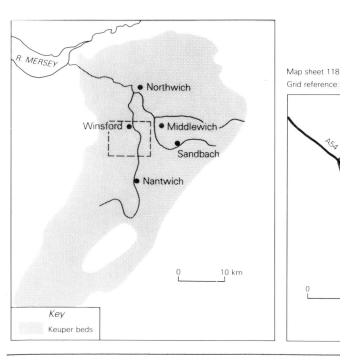

Key

Keuper beds

0 ___ 10 km

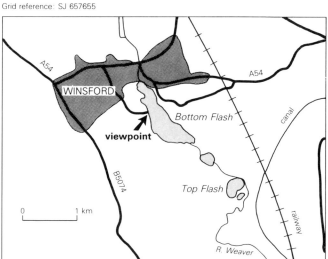

Map sheet 118

Grid reference: SJ 657655

0 ___ 1 km

were full, and the demands upon them slight, there was little fear of serious subsidence. Major subsidence began, however, when the rate of exploitation was speeded up in the early 19th century. The brine reservoirs were soon depleted in the vicinity of the salt works and, as a consequence, the pumps began to attract brine from a greater distance, and long subsurface solutional caverns were carved out by the brine flowing to the pump shafts. It was such a tunnel that collapsed in the 1870s when Winsford Bottom Flash was formed. Thirdly, the Keuper rocks overlying the salt bands are often weak and relatively thin, meaning that collapse takes place rapidly. Finally, the seams of salt are often thicker than seams of coal, so that the amount of subsidence is correspondingly greater.

Modern technological advances

Havoc caused by subsidence in the late 19th century

have effectively eliminated this problem. But, in the 19th century, subsidence was a frightening and sometimes lethal event in the vicinity of Middlewich, Nantwich and Winsford. Houses collapsed, railways undulated, low-lying land became flooded, and canals suddenly drained.

16 The Eglwyseg Screes and the wandering Dee

Llangollen, a town in the borderlands of North Wales, is well known for its International Music Festival, water sports and nearby canal architecture. It also hosts a remarkable variety of natural landscapes, which are not so widely appreciated. In particular, the white Eglwyseg scarp (Creigiau Eglwyseg) stands out boldly, like a shield behind the town and the meanders of the River Dee. The scarp, impressive in its surroundings, is best seen from the steep valleyside just south of Llangollen (for example, from SJ 215410). Indeed, its name echoes the grandeur of the landscape, for it means 'rocks that fell within the domain of the church'.

The existence of the grand scarp reflects the presence of Carboniferous Limestone, a rock whose massive nature and low porosity give it a resistance to physical weathering and erosion not shared by the softer shales and volcanic rocks (of Silurian age) to the west, or by the coal-bearing rocks (The Coal Measures Series of Carboniferous age) to the east. Moreover, the eastward tilt of the Carboniferous rocks exposes their full thickness to the west, thus encouraging the development of a steep scarp face. Closer inspection reveals the scarp towering almost 600 metres above the Silurian rocks, a

wall of bare limestone devouring the rolling, grassy meadows beneath.

This wall consists of an upper, vertical cliff of limestone, with an apron of scree beneath it, as at Wastwater (site 3). The cliff face shows the different layers in the limestone, deposited as sediment in the sea over 300 million years ago. Vertical joints contrast with the layers, the joints resulting from compaction and uplift of the rock. (Uplift also caused the tilting of the layers.) Many joints and planes between the layers (the latter called 'bedding planes') have been widened subsequently by water percolating through the limestone and dissolving

The Eglwyseg screes and scarp

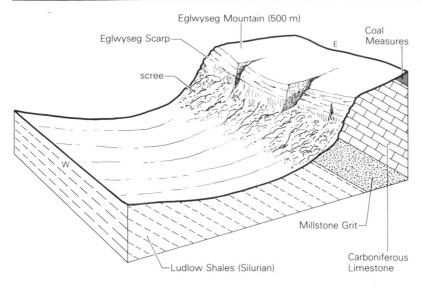

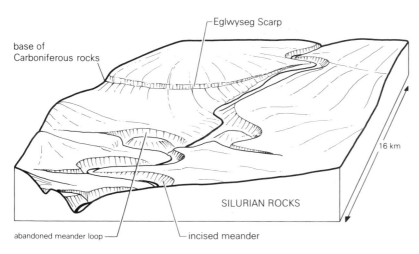

The rock structure at the Eglwyseg scarp (above) and the incised and abandoned meanders of the Dee (below)

some of it in the process. This happens because rain water is acidic and the limestone is alkaline.

The scree has several distinctive characteristics. It is composed of angular pieces of rock that have fallen down from the limestone cliff face above (termed the 'free face'). These angular fragments are often loosened in winter by the action of freezing and thawing of water in the joints of the rock. This prises small fragments, the size of gravel and small boulders, away from the rock face. These collect at the base of the scarp and may, in time, bury most of the cliff face. On the Eglwyseg scarp about two-thirds of the cliff is covered, and therefore protected, by the scree. The scree surface is inclined at more or less the same angle (approximately 30°) along the length of the scarp. This is the highest angle at which the surface of rock fragments is stable; any higher and the fragments will roll down slope under the influence of gravity and, in doing so, the angle of slope will be lowered.

Two other features are also well shown by the Eglwyseg scarp. First, scree is formed more easily at some places than others, due to the jointing in the rock. The cliff face at these points is often slightly recessed, and the scree falls out from here in the form of a fan. Many of these fans then join together to give the apron. In places the apron of scree has been stabilised by patches of grass and the odd bush, whose roots bind the loose fragments. Secondly, small steep valleys, little more than large gulleys, are carved into the scarp face. These can be seen, for example, east of Tan-y-graig (SJ 223453). They have been eroded by running water, possibly when the permeability of the limestones was reduced by frozen water in the pores and joints (permafrost) during the Ice Age. Today some of these valleys are dry. All of the features mentioned above are seen from the road between the villages of Plas-yn-Eglwyseg (SJ 217462) and Tan-y-graig.

To the south of the Eglwyseg screes, the River Dee tumbles through its narrow pastoral vale. The Dee has cut down deeply into the surrounding Silurian rocks, forming beautiful incised meanders. However, it does not meander as much as it used to, for several of them have been abandoned. The Dee has shortened its course in the past by cutting through the neck of two meanders near Llangollen (SJ 220413) and Plas Berwyn (SJ 187434). These abandoned meanders are easily picked out on an Ordnance Survey map (sheet 116) for they appear as broad, horseshoe-shaped valleys lacking major rivers. The fact that glacial debris is found in the abandoned valleys, beneath the patchwork of fields, has led to the suggestion that the shortening took place when the river was swollen by glacial melt water and thus had a greater power of erosion. The old meander cliffs and cores are presently wooded, in contrast to the cultivated meander floors. Both of these abandoned meanders are well seen from the main roads. Further east the river has carved down through the Carboniferous series of rocks near Cefn Mawr, forming still more incised meanders but paying little attention to the geology of the area. Thus, the Dee near Llandudno provides another example of superimposed drainage similar to that of the River Wye near Ross (see site 41 for more details). One part of the incised valley has been 'bridged' by the famous Pont-Cysyllte aqueduct, carrying the canal.

Map sheet 117
Viewpoint grid references: SJ 217462 (Eglwyseg)
SJ 215410 (Llangollen)
Tourist office, hotels etc. in Llangollen

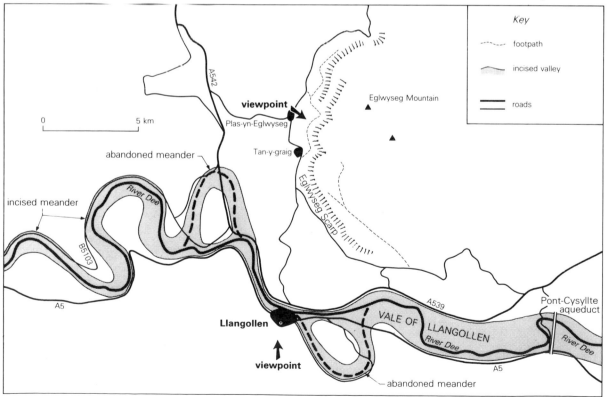

17   Llandudno and its tombolo

In contrast to the grandeur of nearby Snowdonia, the area around Llandudno is subdued. Nevertheless, Llandudno stands on one of the finest examples of a 'tombolo' in Britain. Tombolo is the name given to a coastal strip of land that has grown up between two anchor points, usually the mainland and a nearby island. In general it is composed of mixtures of coastal sediments: clays, sands or occasionally gravels, depending on the local supply of sediments and the energy of the sea. The Llandudno tombolo is composed of clays and sands, probably supplied by the Conwy River and also reworked from earlier glacial debris

deposited in Conwy Bay. (During the Ice Age glaciers extended to the coast of North Wales from Ireland.)

The sheltered and shallow waters in the lee of Great Orme Head encouraged the sea to deposit its sediment. The winds are weak, so wave energy is limited in these sheltered zones and the sea is forced to deposit some of its load. The result is a low-lying area of land linking Great Orme (originally an island) to the mainland. A little relief is provided by hummocks of blown sand, but these are very small in contrast to the steep-sided outcrops of Carboniferous Limestone of the Great and Little Orme. The dense,

massive nature of the limestone (as noted in sites 8 & 13 amongst others) is well seen in the cliffs of the Great Orme.

The Orme's Bay edge of the tombolo, a beautiful curve of sand around which the Edwardians built an attractive crescent of fine houses, has ceased growing. A balanced state appears to have been reached between the amount of beach material supplied and its removal by wave action. It is noticeable that the beach does not extend beyond the limits of protection afforded by the headlands. There is still room for growth on the western side of the tombolo, in the silted river mouth.

Incidentally, a careful look at the Ordnance Survey map (sheet 116) will reveal the old course of the Conwy River, now drained by a small stream. The course can be traced from Llansantffraid Glan Conwy, northeastwards past Dolwyd (SH 817777) and Rhos (SH 830805) to Penrhyn Bay. The present course is thought to have been carved during the Ice Age when the ice extending across the Irish Sea impinged on the coast and blocked the existing drainage. This forced the ponded-up waters of the river to overflow into a new channel.

The breathtaking view of the coastline from Great Orme Head helps to explain why the town of Llandudno developed as a major resort from the middle of the last century. Indeed, the well preserved Edwardian promenade gently following the curve of Orme's Bay could only enhance the natural feature of the tombolo, but subsequent expansion of the town has covered a large part of it. An excellent viewpoint is found on top of Great Orme (SH 774831), just before the precipitous descent into Llandudno. This point is reached via Marine Drive, which begins at the toll to the north of the lifeboat station. The road near Penmaen-bach Point (SH 747786) also affords a good view of the tombolo from over the Conwy estuary, emphasising the low-lying strip of land joining the two Carboniferous Limestone hill masses. The pier in Llandudno provides yet another perspective.

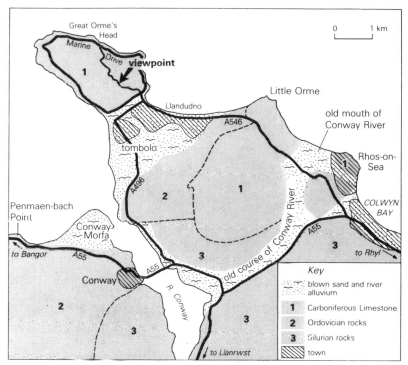

Map sheets 115 and 116
Viewpoint grid reference. SH 774831
Roadside parking at viewpoint

From the viewpoint on Great Orme, looking across the tombolo and bay to Little Orme

18 Nant Ffrancon: the Ice Age's legacy

Telford's London–Holyhead road, the A5, passes along one of the grandest highland valleys in Britain and one of the few mountain passes in Snowdonia – Nant Ffrancon, the deep flat-floored trough of the River Ogwen. In 1862, George Borrow made his descent from Llyn Ogwen and relates,

'At length we came to a gloomy-looking valley trending due north ... having an enormous wall of rocks on its right and a precipitous hollow on the left.'

We can assume that the weather was kinder to Thomas Pennant, whose description in 1773 remains a fine testimonial:

'I do assure the traveller who delights in wild nature, that a visit to ... Nant Ffrancon from Bangor will not be repented. The waters of five lakes dart down the precipice of the middle (of the Ogwen step) and form the torrent of the Ogwen ... This bottom is surrounded with mountains of stupendous height, mostly precipitous; the tops of many edged with pointed rocks.'

The floor of Nant Ffrancon is more than 750 m below the level of the enclosing ridges, yet they are no more than 1500 m away. The main valley has a classic U-shaped cross profile, a product of Ice Age glacial scour, that contrasts with the V-shaped profile more characteristic of normal river valleys.

On the western flank of the valley small glaciers excavated nine hollows called, in Welsh, 'cwms' (Scottish: corrie; French: cirque). Similar landforms are not found on the west-facing slopes of Penyrole-wen. This is probably explained by the climate. The slopes beneath the Glyder plateau would be ideal sites for the accumulation of wind-driven snow coming from the west, and being north-east facing they would receive much less heat from the sun. This combination would provide opportunities for glaciers to accumulate, and for the cwms to grow as a result.

The sequence of events is as follows. First of all, during the maximum extent of glaciation a great ice cap to the east of

Snowdonia, called the Merioneth Ice Cap, radiated outwards from an ice dome at least 650 m thick. This sheet of ice overran the mountains in its path and gouged out deep U-shaped valleys along lines of weakness in the rocks, of which Nant Ffrancon is one. The highest peaks, such as Penyrole-wen, stood up as islands above the ice, called nunataks. These peaks were not subjected to glacial erosion, but intensive frost shattering took place on them, leaving their surfaces angular and jagged – a marked contrast to the glacially smoothed areas below.

The second stage of development took place before and after the maximum development of the Merioneth Ice Cap. Because they were higher, the mountains maintained small glaciers even though the ice cap was not present. It was these that enlarged the hollows on the west side of the valley, by plucking and grinding at the rocks, to produce the cwms, the most spectacular of these being Cwm Idwal. (See site 4 for further details about cwm formation.)

The third stage of development took place after the glaciers finally retreated 10000 years ago. It was then that the landscape we see today was born. In the floor of Cwm Idwal, for example, a large lake formed (Llyn Idwal), fed by water tumbling down the gullies that dissect the precipitous back and side walls of the cwm. A small mound of glacial debris (a moraine), dumped at the edge of the glacier as it melted, helps to dam the lake.

Besides the U-shaped valley of Nant Ffrancon and the cirques that dig into its western flank, there is other evidence of glacial erosion. Just above Rhaeadr Ogwen, an accessible and striking vantage point for the whole landscape, is a feature called a 'roche moutonée'. This is a knob of bedrock over which the glacier has passed. The side of the rock facing the flow is rounded and often covered in deep scratches called 'striations'. These are caused by the

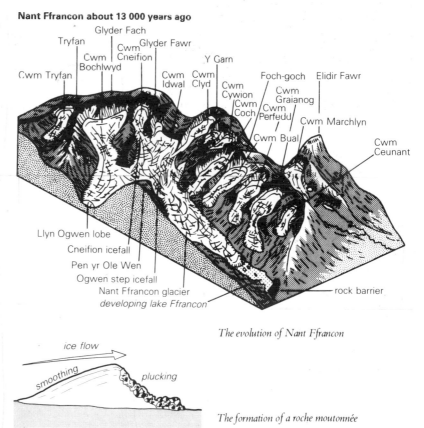

Nant Ffrancon at maximum extent of Merioneth ice cap

high-level Glyder cirques
central Glyder nunatak
Idwal icefall
developing trough of Cwm Idwal
Tryfan nunatak
Llanberis outflow glacier
northern Glyder nunatak
Elidir Fawr
Llanberis outflow glacier
Ogwen step icefall
Pen yr Ole Wen nunatak
Nant Ffrancon glacier

Nant Ffrancon about 13 000 years ago

Glyder Fach
Tryfan
Cwm Cneifion
Glyder Fawr
Cwm Bochlwyd
Y Garn
Cwm Tryfan
Cwm Idwal
Cwm Clyd
Foch-goch
Elidir Fawr
Cwm Cywion
Cwm Graianog
Cwm Coch
Cwm Perfedd
Cwm Marchlyn
Cwm Ceunant
Cwm Bual
Llyn Ogwen lobe
Cneifion icefall
Pen yr Ole Wen
Ogwen step icefall
Nant Ffrancon glacier
developing lake Ffrancon
rock barrier

The evolution of Nant Ffrancon

ice flow
smoothing
plucking

The formation of a roche moutonnée

ice dragging other rocks embedded in its sole over the obstacle in its path. The down glacier side of the roche moutonée is steep, jagged and broken. The ice would have plucked pieces off it in passing. The name is a curious one: it was used in the late 18th century by a French-speaking scientist, called Saussure, who described the rippled, glistening effect produced by a whole series of them in the Alps as 'roches moutonées', in fancied resemblance to contemporary wigs slicked down with mutton tallow!

The best views of the valley and Cwm Idwal are from near the path that starts at the Mountain Rescue Post (SH 649604) and proceeds, across the moraine at the edge of the cwm that dams the lake (at SH 647599), to Llyn Idwal itself. The more adventurous can then scramble up through the boulders and foxgloves and along the narrow ridge, called an 'arête', at the top of the backwall of the cwm. This route circles around the Devil's Kitchen and crosses the glacially smoothed col (near Llyn-y-cwn at SH 636585), across which the Llanberis Pass and valley can be reached. If you intend to go hiking in the mountains please ensure that you are well prepared and equipped.

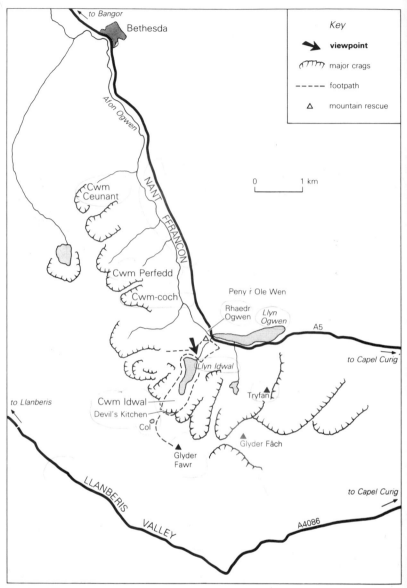

Map sheet 115
Viewpoint grid reference: SH 646600
Parking on roadside

19 Llyn Llydaw: an imprisoned lake

One of the greatest scientific discoveries of the first half of the 19th century was that the Earth had undergone an Ice Age. Many of the earliest confirmations of this fact were made in the Alpine valleys of Switzerland, but it did not take long for British geologists to scour the highland areas of Wales for similar evidence. One such scientist was A. C. Ramsay, and his investigations took him to Llyn Llydaw in Snowdonia:

'Approaching Llyn Llydaw, the full grandeur of this wonderful valley bursts on the beholder. A lake rather more than a mile in length and of a green colour, like some of the lakes of Switzerland, obliquely crosses the valley. Around it rise the cliffs of Lliwedd, Cribgoch, and Pen

Wyddfa, seamed with veins of white quartz that gleam like streaks of snow on the tall black rocks circling the vast amphitheatre, the scarred sides and ragged outlines of which, sharply defined against the sky, may well seem, till attempted, hopelessly inaccessible to the unpractised climber. In every season and phase of weather, there is a charm in this valley to the lover of the mountains … The signs of a glacier are so evident in Cwm Llydaw that it is needless to describe all the details.'

The easiest way to approach Llyn Llydaw today is along the small track from Pen-y-pass (SH 647556), itself on the scenic road through the Llanberis Pass. The track takes you first to Llyn Teyrn (SH 642547), a small lake occupying a glacially

scoured hollow in the tough volcanic rocks of Ordovician age. A close inspection of some of the rock outcrops near Llyn Teyrn will reveal that the volcanic rocks appear to be made up of a multitude of small hexagonal columns placed side by side. This columnar jointing resulted from the cooling and contraction of the molten material as the rock formed, and it is typical of certain volcanic rocks. Much of the rock surface was rounded and smoothed by the glaciers passing over it during the Ice Age, but in Postglacial times frost shattering has littered the area with small piles of coarse, angular fragments of the dull, grey rock.

A more spectacular example of coarse, angular scree is found beyond Llyn Teyrn, where the debris coats the hillside beside the footpath. The

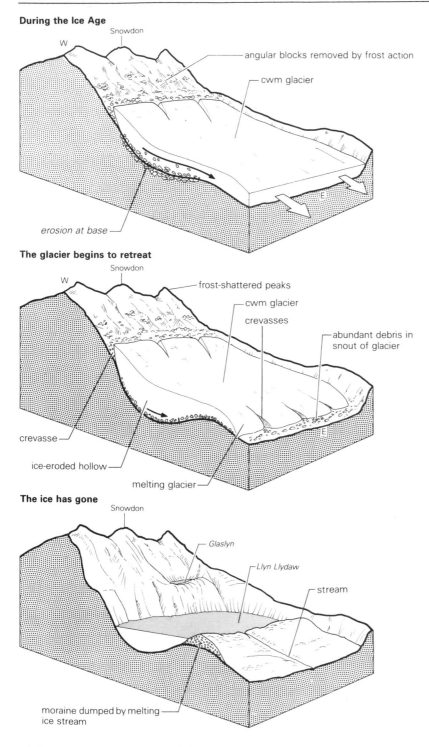

During the Ice Age

Snowdon

W

angular blocks removed by frost action

cwm glacier

E

erosion at base

The glacier begins to retreat

Snowdon

W

frost-shattered peaks

cwm glacier

crevasses

abundant debris in snout of glacier

crevasse

ice-eroded hollow

melting glacier

E

The ice has gone

Snowdon

Glaslyn

Llyn Llydaw

stream

moraine dumped by melting ice stream

The formation of the moraines at Llyn Llydaw

frost-shattered blocks have been supplied in this case from exposed rock faces higher up the hillslope. Evidence of earlier glacial action is also to be seen nearby, in the form of striations on the smoothed volcanic bedrock. Striations are the scars left by stones and boulders in the glacial ice as they are dragged across the bedrock surface. These scars vary from mere scratches to gouges over a centimetre deep and they provide useful indicators of the direction of ice movement in the area. Often the differing trends of the striations, the columnar joints, and the brilliant white quartzite veins that cut through the rocks, give the landscape a chaotic appearance. These linear and angular features contrast markedly with the glacially smoothed rock surfaces, plate-like patches of lichen and rounded glacial boulders.

The last bend in the path reveals Llyn Llydaw, a large lake occupying a cwm and transformed in recent years into a reservoir. The cwm was gouged out by snow and glacial ice, and is dammed very effectively by a natural barrier comprising a series of hummocky mounds rising up to 5 m above the level of the lake. These mounds rest on the rock lip of the cwm, and are termed moraines. The unsorted mixture of silt, sand and gravel-sized debris making up the moraines can be seen in the exposed surfaces nearest the lake (SH 635545). It is likely that the debris was dumped at the foot of the glacier occupying the cwm when the glacier melted, thereby sealing the cwm lake. These moraines are thought to have formed in a glacial phase over 20000 years ago, and they are some of the finest examples of such features in Snowdonia.

The other characteristic features of the cwm are similar to those described in the Lake District (site 4) and elsewhere in Snowdonia (site 18). The forbidding cliff walls of the cwm, which plunge spectacularly over 300 m from the razor-edge arêtes into the cold waters of the lake

21   Snowdonia's valley lakes

Snowdonia is a land of lakes. Broadly speaking they fall into two categories: cwms and valley lakes. We discuss the former in the Lake District (site 4) and elsewhere in Snowdonia (site 19), whereas the valley lakes have been sadly neglected, considering their prominence in the landscape.

Lakes form where there is a dip in a river valley, like a hollow, which fills up until the water level is just higher than the lowest level of the surrounding land, whereupon the water flows out, down stream. This hollow can be created by erosion of the bedrock, or by deposition of debris effectively forming a dam, with a false hollow created behind it. Sometimes both erosion and deposition are involved.

The erosional lake basins are encouraged to form under three sets of conditions: first, where the rock types are less resistant and thus more easily eroded than the surrounding rocks; secondly, where the rocks have been structurally weakened, for example, along a fault line where there are zones of shattered rock on either side of the fault; thirdly, where ice or water erosion is concentrated in a particular area, maybe beneath a waterfall or down stream of a constriction in the flow of ice. Each of these circumstances can result in the hollowing out of bedrock.

Hollows resulting from deposition are often more transient: that is, they last only as long as it takes the river to erode through the depositional dam. Depositional hollows can be

formed in a variety of ways. Melting of an ice sheet or glacier will leave behind it an uneven sheet of debris, which may contain small hollows later filled by melt water or rain water. More specific damming across a valley results from deposition of a glacial moraine during the melting of a glacier. Moraine dams are common in Snowdonia and the Lake District, both for valley lakes and corrie lakes. Damming can also be caused by movements of masses of land, such as in a landslide or mudflow. Thus, there are a variety of ways in which lakes of varying size, age and permanency can be formed.

The lakes of Snowdonia illustrate these differing origins very well, although the origin is still disputed in some cases. Take, for example, Llyn

Gwynant (SH 645520), which occupies a glacially scoured hollow in the U-shaped Nant Gwynant valley and is dammed by moraine. Incidentally, a very good example of a 'roche moutonée' (see site 18) is to be found near the north-east shores of the lake, and can be well observed from the A498.

A more complex example of lake evolution is offered by Llyn Ogwen at the head of Nant Ffrancon (SH 655604). Here, the glacially scoured lake originally drained to the east into the Afon Llugwy, the drainage divide being at this time at the Mountain Rescue Post (SH 649604). It is thought that subsequent glacial erosion lowered and smoothed the drainage divide so much that this end of the lake became lower than the eastern end. (At the same time the lake may have been deepened or even formed.) The drainage divide was not smoothed out completely, so that a rock step remains near the western edge of the lake, and water overflows down this into Nant Ffrancon.

Beautiful Lake Bala (SH 910340) in southern Snowdonia provides an example of the exploitation of structural weakness in the formation of a lake. The lake sits along a fault zone and has been carved out of shattered and, therefore, easily eroded rocks alongside the fault. It is difficult to imagine this when sharing the tranquility of the place today, either from the roadside (A494, B4403) or the narrow-gauge railway that runs along the eastern shore of the lake. It is likely that glacial action was responsible for the erosion of the weakened rocks, as in the other examples, but the trend of the lake is definitely controlled by the fault line. The movement along the fault is horizontal (rather than vertical) and so it is termed a transverse fault. Although movement is not occurring today, there is evidence of a horizontal shift of over 3 km having taken place well before the beginning of the Ice Age.

Tal-y-Llyn (SH 720100) to the south-west is developed in weakened rocks associated with the same fault line as Bala Lake. Here, however, the lake has also been dammed by the deposition of a great mound of debris (as seen just behind the hotel; SH 710094). There is some dispute about the origin of this mound, some believing it was a moraine, whereas others maintain that the coarse and poorly sorted debris is part of a rockfall. Either way, the lake has been effectively dammed, providing pleasure for windsurfing enthusiasts.

The valley lakes and neighbouring sites in North Wales

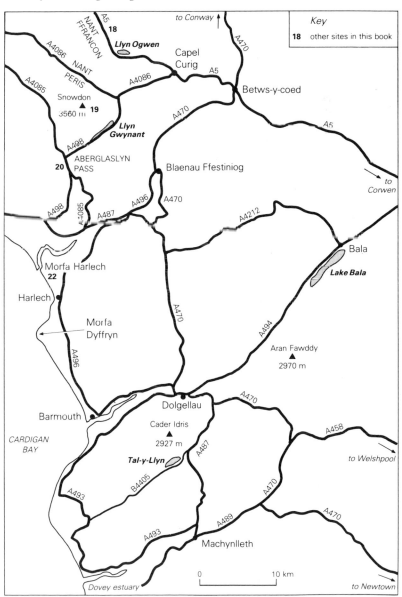

Map sheets 115, 124, 125

22 Harlech Spit and the guarded cliff

The castle at Harlech 'standeth advanced on a very steep rock, and looketh down to the sea from aloft, which being built as the inhabitants report, by King Edward the First, took name of the situation. For Arlech in the British tongue signifieth as much, as upon a stony rock'. So wrote Saxton in the 17th century when describing the castle, built between 1283 and 1290 as a garrison for the English. The existence of a water gate at the castle shows that, when it was built, the sea lapped at the edges of the rock face which provided an ideal defensive position.

Today the castle (at SH 581312) is well inland, separated from the Glaslyn estuary by nearly 5 km of sand and marsh, all of which appears to have accumulated within the past 700 years. The castle is built on a steep rock face that extends in a northeasterly direction to Talsarnau (SH 612360) and Bryn Glas (SH 623375) beyond. This is the old cliffline, washed by the sea in the recent past, and affording views to the north of Snowdonia and the Aberglaslyn estuary. It is best seen from the castle itself. The old cliffline can also be distinguished to the north of the present Glaslyn estuary, where it is roughly outlined by the A498 and A487. It is possible that the sea once extended as far inland as the Aberglaslyn Gorge (site 20), but much of the former muddy estuary north and east of Porthmadoc was enclosed and drained in the early 19th century. Part of the Harlech salt marshes were enclosed at the same time, the embankments being visible from the A496 near Talsarnau. These estuaries of the rivers Glaslyn and Dwyryd are former glaciated valleys, drowned by the rise in sea level at the end of the Ice Age. Similar features are described in sites 26 & 64.

Morfa ('marsh' in Welsh) Harlech has accumulated as a triangle of land against the Harlech cliffline, growing out across part of the Glaslyn estuary. It is a classic example of a sand spit with salt marsh growth behind it. The existence of the spit is due to three factors, namely the prevailing southwesterly winds, the shape of the coastline, and the supply of sediment. Sand is supplied from glacial debris offshore and by longshore drift from

66

the south, and the finer sediments are carried down to the estuary by the combined efforts of the Glaslyn, Cynfal and Dwyryd rivers. The southwesterly wind is responsible for creating waves that move sand northwards along the coastline and inland from offshore deposits. This makes it possible for the sand spit, composed of beach sand and dunes, to grow northwards from Llanfair, where the fossil cliffline changes direction so that it is no longer at right angles to the prevailing southwesterly wind.

Growth of the sand spit, like a bar attached to land at one end, is also encouraged by the protection provided by the coastline to the north of the Glaslyn estuary; that is, the spit is not growing out into open sea. However, the spit will never be able to join onto the northern coastline, since a drainage channel will be needed for the rivers flowing into the estuary from Snowdonia. Thus, the sand spit at Harlech has grown almost as far as it can without the help of man in further embanking and artificially draining the area.

It is noticeable that the seaward edge of the spit has grown at right angles to the prevailing wind and to the waves. This angle is created by the waves to allow efficient movement of sand along the coastline by longshore drift. The edge of the spit is therefore in balance with the prevailing wind, and, through the wind's effect on the sea, the sediment supply. The spit protects the area inland, thereby reducing the energy available for erosion, and this encourages the deposition of sand and clay brought down by the rivers. In this way a low-lying marshland accumulates behind the sand spit, a salt marsh that is crossed by muddy drainage creeks and is inundated by the sea at every high tide. The embanked area is no longer flooded and it provides poor pasture for sheep.

The best views of the area are obtained from the A496, which follows the old cliffline and provides

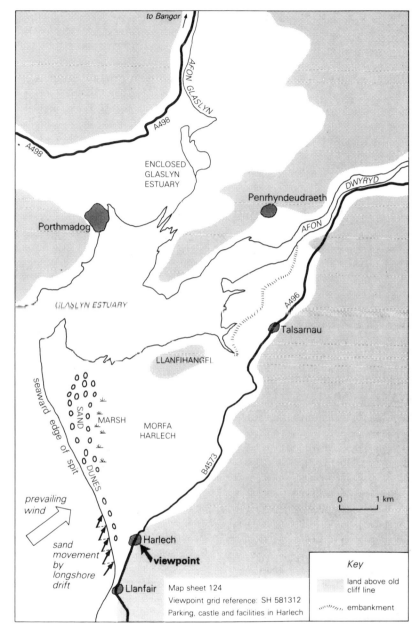

vistas of the marsh interspersed by rocky hills that once formed islands in the sea. Some of these hills have small amounts of windblown sand banked against their south-west facing slopes. The sand spit with its partly vegetated dunes is best viewed from the castle or nearby (SH 579307).

Similar spits are found to the south near Llanbedr (Morfa Dyffryn), Barmouth (Ro Wen), Aberdovey and Borth. All of these have formed in response to the prevailing wind, although Morfa Dyffryn has grown out to a moraine 'island' and is, thus, aligned in a somewhat different direction to the others.

23 Stiperstones: the Devil's tors

The Stiperstones ridge of Shropshire has an imposing skyline. Groups of craggy grey tors rise from the steep slopes that run through heather, bracken and pine to the lusher land in surrounding valleys. Many local legends have arisen to explain their origin, legends involving witches and the Devil. It is indeed remarkable how in the mythology of British folklore so many classic landforms have been seen as the works of Satan!

For example, the highest tor, which rises 10–20 m above the general level of the ridge, is named the Devil's Chair. According to one story the Devil came to Shropshire from Ireland with an apron full of stones in order to fill up Hell Gutter, which lies on the side of the hill. He sat down for a rest, but, when he got up, his apron string broke and all the stones were scattered around the

Devil's Chair, where they remain to this day. A second story has it that the Devil hates England above all the other countries in the world, and because he believes that, when the Stiperstones sink into the Earth, England will perish, he visits from time to time and flings himself down in the chair in the hope that the weight will sink the hill. Scientists have other explanations for the ridge, its tors, and the angular debris that lies about them.

The ridge, formed of tough Ordovician quartzites, runs from Lordshill Valley (SJ 385020) through the Hollies, Blackmoorgate, the Paddock, Shepherd's Rock, Scattered Rock, The Devil's Chair, Manstone Rock, Cranberry Rock, and Nipstone Rock, to the Rock (SO 351962). Its general trend is north-east to south-west, and it reaches a height of about

525 m above sea level in the vicinity of Manstone Rock, and rises some 150–180 m above the level of the surrounding valley bottoms, most of which are in less resistant shales. The quartzite which makes up both the ridge and the tors is a hard, white, sandstone with layers of conglomerate and occasional thin shale bands. The rock is moderately to highly jointed and, where exposed in the tors, it can be seen to tilt at high angles, generally to the north-west. The tilt, associated with the presence of faults, accounts for the overall form of the Stiperstones escarpment.

The tors mostly outcrop along the crest of the ridge, and their very steep sides (often in excess of 50°) contrast greatly with the gentler (about 10°), smooth slopes of the ridge. Varying rates of rock breakdown and erosion at different places along the ridge have revealed the tors. (Similar tor formation is described in more detail in site 60.) Of particular interest are the piles of angular boulders, the debris left from the weathering of the jointed quartzite, surrounding the tors.

A close scrutiny will reveal that the boulders have been sorted into remarkably clear patterns of stripes and polygons, arrangements of stone with heather and bilberry in between. Although the width of the stripes varies, from 1 m to almost 10 m, they are still strikingly regular in their form. The rock stripes average just over 3 m in width, and the vegetation stripes in between average slightly less than 3 m. The individual boulders alone may be as large as 2.5 m in diameter, though the average is about 0.5 m. Moreover, the boulders are not oriented randomly; often their longest dimension points down slope, and individual slabs are commonly arranged on edge rather than flat.

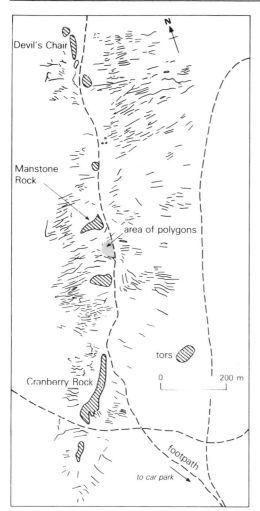

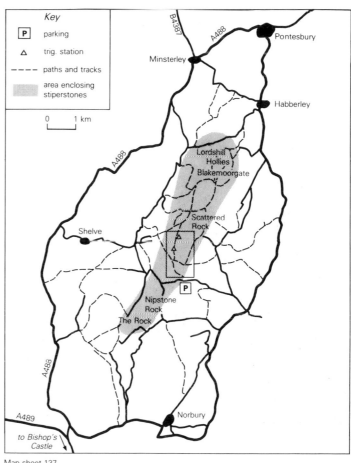

Map sheet 137

Manstone Rock grid reference: SO 367987

The polygons are best developed on flat and low-angled ground, but they also occur on slopes of moderate steepness (up to 7.5°). Higher slope angles favour the organisation of stones into elongated polygon patterns, whereas stripes are predominant on even steeper slopes of up to 16°. Thus, the polygons on the crest of the ridge often lead down slope to elongated polygons and then to stripes. However, polygons are of more irregular shape (average diameters of 7–9 m) and limited distribution than the stripes.

The large spreads of boulders forming the stripes and polygons have plainly been derived from the ridge of quartzite, of which the tors form isolated remnants. In fact, the patterned ground gives an important clue to the origin of the tors, for today such large-scale patterned ground forms under conditions of extreme cold, in such areas as Alaska and northern Scandinavia. Thus, intense frost activity probably accounted for the removal of the angular boulders from the ridge, leaving the tors as mere remnants of a formerly larger ridge.

The heavy cover of lichen on the tors and on the stripes indicates that frost action is relatively insignificant today. We must look back to the Ice Age, when valleys round the ridge were glaciated and the area was on the margins of a great ice cap, to explain the formation of the tors and patterned ground. It was perhaps around 18 000 years ago that these features formed under such conditions.

The most striking polygons occur on the crest of the ridge between Cranberry Rock (SO 366981) and the Manstone (SO 367987), and also to the south of The Devil's Chair (SO 369991). Others can be seen on the crest of the ridge in the area around the Paddock (SJ 377006). Stripes, however, can be seen around most of the major tors.

24 Piracy at the Devil's Bridge

The crashing waterfalls, steamy subtropical atmosphere, dense undergrowth and spectacular gorges attract people from far and wide to the Devil's Bridge, one of the most breathtaking landforms in Wales, albeit a commercially exploited one. Despite its long popularity the area remains largely unspoilt by man and is a testimony to the forces of nature.

Two major sets of waterfalls are seen near Devil's Bridge, the first created by the River Rheidol (Gyfarllwyd Falls) as it turns sharply towards Aberystwyth. The second is the Mynach Falls, where the river tumbles through a narrow chasm into the lower, coastal part of the Rheidol valley below. One swirlpool in the upper course of the Mynach falls is a superb example of the hollowing effect of turbulent water and boulders even on relatively resistant rocks. The resulting rock surface is smoother than the average bathtub. The air all around the swirling water is heavy with water vapour and droplets, providing an almost steamy atmosphere in the shady, dank chasm. In this atmosphere mosses, creepers and ferns flourish, giving a jungle-like feel to the place and a rich peaty smell. A glance upwards will reveal the original bridge over the chasm, the Devil's Bridge.

Two newer bridges have since been built above this, and these, in addition to the lower part of the Mynach Falls, can be seen during the longer walk from the Devil's Bridge down to the base of the falls, and up to the hotel on the other side of the gorge. This section of the gorge is more open, allowing light to penetrate and to disperse the gloom seen in the swirlpool section. Likewise, the vegetation appears less dense, with oaks and rhododendrons as well as ferns. The river descends to the valley below in a series of four or five major, frothing falls,

interspersed with rock pools holding swirling, peaty-brown, acidic water.

The existence of the gorges can be fairly easily explained by 'river capture'. Before capture, the Rheidol flowed southwards from Ponterwyd, through the Devil's Bridge area, and continued southwards. The Mynach river probably joined the southward-flowing Rheidol near Devil's Bridge. At this time the lower, westward-flowing part of the present Rheidol was a much smaller, independent river flowing towards Aberystwyth, as today, but the headwaters did not extend as far east as Devil's Bridge. Rapid erosion, downwards and headwards (that is, the headwaters of the stream eating backwards) by this lower section of the Rheidol brought the headwaters nearer to the southward-flowing upper Rheidol and Mynach rivers, which were unable to erode so rapidly and therefore occupied valleys at higher

levels. Eventually the lower Rheidol eroded so far backwards that it cut through the higher valley, and thereby diverted the drainage of the upper Rheidol to the west. Capture of the Mynach river may have taken place at the same time as that of the Rheidol, or slightly later, aided by further headward erosion. The capture of water, from higher valley systems, explains the existence of the waterfalls and the sharp change in direction of the Rheidol river (termed the 'elbow of capture').

The rapid change in gradient at the falls has given increased erosive power to the rivers, and the Rheidol has responded by incising a deep valley into the rocks just above the falls. Eventually this may lead to the smoothing out of the long profile of the valley, and the lengthening of the gorge.

Excellent overviews of the Rheidol gorge and the capture are to

be seen from the hotel (SN 740770) and from the B 4343 just south of Devil's Bridge, but detailed examination of the falls is possible only from the commercial pathways which start at the bridge itself (SN 742770).

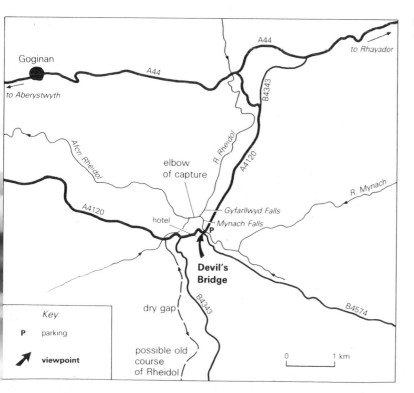

Map sheet 135
Grid reference: SN 740770

25   Tregaron Bog: an overgrown lake

Tregaron, a small village in mid-Wales, has the distinction of being perched beside the biggest set of bogs in Britain. The three great bogs together cover an area of over 15 km² in the upper reaches of the Teifi valley. Follow the Teifi river up stream of the bogs and you will enter an area of upland so remote that it has been called the 'Teifi Desert'.

Within this bleak landscape the bog is a rather gentle feature – a sea of grasses, sedges and reeds dissected by drainage channels, streams and the meandering River Teifi. It provides shelter and home for a profusion of wildlife, including herons and hawks, and the rare black adder is also believed to inhabit the area. Cotton grass, vetches and trefoils flourish in the peaty soils, but the rich flora and fauna are strictly preserved by the Nature Conservancy Council.

Borings into the bog have revealed its origin and history. At the base, over 8 m below the surface, there is a thick layer of blue–grey clay, laid down in the still waters of a lake when the surrounding land was still cold and unvegetated. This layer passes upwards into brown coloured muds containing seeds of open-water plants and fragments of wood. Above this are various layers of muds and peat; accumulations of dead organic matter formed as the lake was increasingly filled in with mud and invaded by plants that grew out across the floor of the lake, and subsequently died leaving the peat. Today, all the old lake has been filled in by mud and peat layers, leaving a classic acid peat bog covered by a variety of heathers, mosses, deer-grass, club-rushes, cotton grass, and the beautiful and protected bog asphodels.

To appreciate the former existence of a lake you need a good vantage point, such as at Pont Einon (SN 672613) or Pen-bryn (SN 684648), and the Ordnance Survey map (sheets 135 & 146). Looking out over the flat bog, little imagination is needed to see it as a large lake fed by the River Teifi. The origin of the lake can be unravelled with a careful look at the map, which shows a strip of higher land enclosing the southern edge of the bog. The River Teifi has carved through this ridge, which formerly extended across the valley. The ridge is a glacial ridge (moraine) left as the ice retreated from the area at the end of the Ice Age. It provided an effective dam for the river, thus forming the lake behind it, until the level of the impounded water was such that the river could overflow the moraine at its lowest point and carve a channel through it. Beyond the moraine, the river flows across

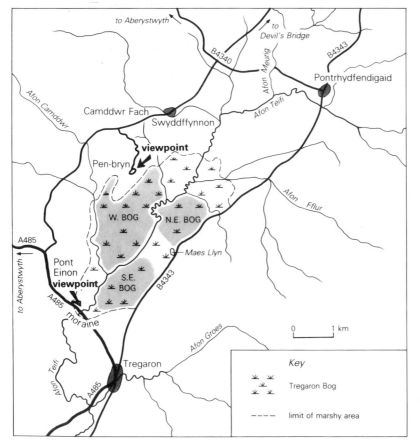

Map sheets 135 and 146

Viewpoint grid references: Pen-bryn SN 684648
Pont Einon SN 672613
Roadside parking

outwash gravels deposited by streams and melt water that flowed out from the ice sheet.

Over the past 10 000 years or so the River Teifi has gradually eroded down through the moraine, lowering the level of the outlet to the lake. As a result, much of the water drained out of the lake. Eventually, the combination of drainage, deposition in the lake of mud brought in by the streams, and the formation of layers of peat has created the present bog.

26   The drowning of Solva Harbour

Like so many of the British landforms, Solva Harbour has been carved out during the Ice Age. The area around the harbour is made up of a series of hardened Cambrian and Ordovician rocks, which were planed down by the action of the sea about 10 million years ago. The resulting coastal platform now stands at approximately 60 m above the sea, and as such provides a useful bench from which to view the huddled cottages and sheltered boats of the small harbour, and to admire the natural landscape.

An especially good viewpoint is to be found on the dolerite (of Ordovician age) headland to the south-east of the harbour. After crossing the small stream at the mouth of the valley (SM 806243), climb to the viewpoint (SM 803239) along any one of many narrow paths. After this climb, the excellent defensive position of the harbour is obvious. But you would not have been the first to think of defence, for Iron Age man occupied this headland, and a careful look around will confirm this. Even as late as the 19th century, Solva was a thriving port.

What is Solva Harbour? From the viewpoint you will notice a similar feature on the other side of the headland. This is the Gribin Valley. However, it has been infilled to a greater extent than the Solva Valley, by clays, silts and sands deposited by

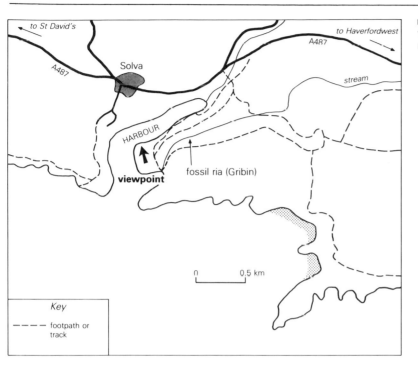

Map sheet 157
Viewpoint grid reference: SM 803239
National Trust and village for parking

A bird's eye view of Solva Ria, showing infilling by siltation at its head

the river and the sea. These two inlets are drowned river valleys ('rias'). The valleys were carved into the coastal platform, and their lower portions were drowned subsequently by a rise in sea level. Since then a small beach has formed around the shoreline. If you look up stream you can see the meandering course of the valley, and its partial infilling with river muds (termed alluvium) that has resulted in a flattened valley floor: a depositional U-shaped valley.

Two problems immediately spring to mind concerning the origin of the rias: first, the conditions under which the gorge-like valleys were carved; and secondly, the date of drowning.

That the Solva Valley was formed by river rather than marine or glacial action can be shown by its meandering nature and original V-shaped cross section (ignoring the relatively recent deposits filling the floor). However, it was probably gouged under conditions of higher stream flow than at present, because the small stream in the valley does not have the necessary power to carve such large valley meanders. The greater stream flow may have resulted from changing climate, possibly during an interglacial phase, with much rainfall and water runoff from the land surface. Alternatively, meltwater streams might have carved the valleys, issuing from, or flowing under, great ice sheets that covered the area during one of the glacial phases. The latter explanation is more likely since conditions during interglacials are believed to have been very similar to those at present. Moreover, the form of that part of the valley presently drowned by the sea can be explained only if the sea level was lower than that today, and sea levels were much lower during the glacial phases.

At the close of the last glacial phase approximately 12 000 years ago, sea level rose gradually as melting ice sheets returned water to the oceans. Dates obtained from the remains of a drowned forest in Freshwater Bay (a few kilometres to the south-east) suggest that the sea in the area reached its present level about 6000 years ago. This date marks the last drowning of the ria, although it could have existed during earlier interglacial phases of high sea level, for it is not clear when, in the Ice Age, the initial valley was formed (see site 64 for further details).

Modern-day processes, such as silting of the valleys and beach construction continue to decorate the mould of Solva Harbour. The village remains as a sleepy reminder to the prosperity of the past in the same way that the small stream belies the forces of rushing water that once carved out the landscape.

27 The Green Bridge, Stack Rocks and the Devil's Cauldron

The Pembrokeshire Coast National Park in Dyfed contains some of the most pleasing coastal scenery in the British Isles, with deep inlets, coastal sand dunes, enticing coves and serrated cliffs. Although the limestone, which forms some of the best cliff scenery, outcrops widely between Penally (118992) and Angle Bay (880020), an especially fascinating area occurs on the edge of the Castlemartin Firing Range. This area is easily reached – when military activities permit – by a good road from the crossroads west of Castlemartin Camp. Many people come to this spot to watch birds, for in the spring the dangerous cliffs hereabouts are used by large nesting colonies of guillemots, kittiwakes, razorbills, fulmars and shags.

The geology of the area is dominated by an unusually thick sequence of Carboniferous Limestone (containing some cherty material) which is disturbed by a number of small folds and faults. Storm waves from the south-west batter the limestone cliffs throughout the year and exploit the lines of weakness; some of the layers of limestone are weaker than others, and faults and upfolds are often sites of weakness. This erosion is due to the physical impact and pressure created by the surging waves and the sand carried within them, and by chemical solution of the limestone. Such erosion has led to the formation of many spectacular landforms.

The Green Bridge of Wales has resulted from erosion on both sides of a headland, forming caves which were gradually enlarged back into the headland, until finally they met, giving rise to an arch. The arch is about 24 m high and has a span of more than 20 m. Its grassy surface runs down from the clifftop and its outer support rests on a broad rock pedestal which demonstrates the landward dip of the limestone beds.

To the east of the Green Bridge are two imposing rocks that have become isolated from the cliff by wave erosion. Such isolated masses are called 'stacks'. The Stack Rocks (also called Elegug Stacks) represent the final stage of cliff disintegration, when the arch has collapsed owing to continued erosion, leaving the supporting pedestals as precipitous isolated blocks. The stacks in the course of time will be undermined by the sea, becoming mere stumps of rock. The higher of the two stacks is 32 m tall and probably both it and its smaller neighbour have been formed from the destruction of an old

Characteristic coastal features produced by the marine erosion of limestone

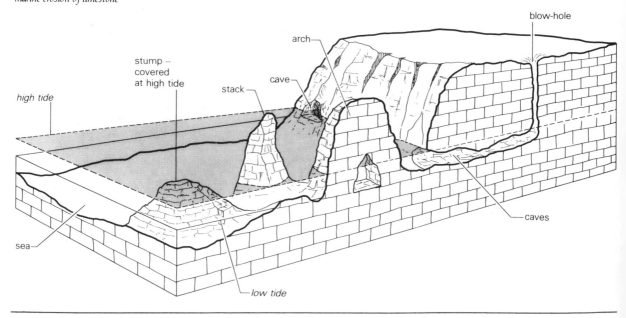

The Green Bridge of Wales

headland which once contained natural arches. Not all stacks, however, develop from arches. Some have a less spectacular history when joints in a limestone headland are widened by the sea and percolating rain water, thus separating a mass from the rest of the headland.

Striking though these features are, the most extraordinary landform on this stretch of coast is the Devil's Cauldron. This is an enclosed shaft some 45 m deep and up to 55 m across. It has developed in the Devil's Cauldron Peninsula, a highly faulted mass of limestone bounded by high cliffs. The shaft has been formed by the growth, coalescence and collapse of a number of blow-hole caves developed along fault-lines. Blow-holes are formed when cave erosion penetrates deeply back into the limestone and, as a result of water pressure caused by wave movement, erosion takes place upwards along lines of weakness. Eventually this upward erosion reaches the top of the cliff and the cave at sea level is linked to the cliff top by a narrow shaft, the 'blow-hole'. On its southern side the Cauldron is connected with the open sea by a bridge 18 m high, with a span of 21 m. In all it has been calculated that, to produce the Cauldron, limestone solution and cave collapse have removed a weight of rock heavier than the combined mass of the three largest ocean liners ever built. As erosion proceeds even further, the Cauldron will break down to form a series of stacks and arches that will dwarf its nonetheless imposing neighbours: the Green Bridge of Wales and the Stack Rocks.

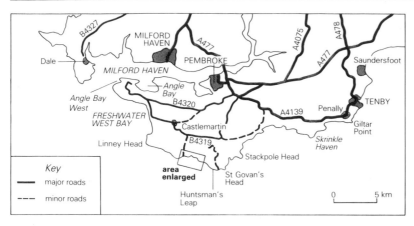

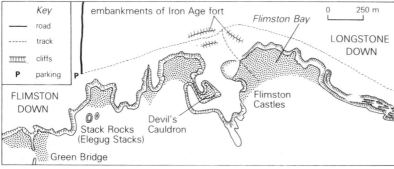

Map sheet 158

Devil's Cauldron grid reference: SM 930945

Many such features can be seen in the area, for example, near Skrinkle Haven (SR 080972) where there is an impressive arch eroded out of vertically dipping limestone. Caves and deep, narrow fissures are ubiquitous. The best example of a fissure formed by erosion along a fault line is the Huntsman's Leap (SR 963929), and caves are especially common near Skrinkle Head and between St Govan's Head (SR 975927) and Stack Rocks (SR 926945). On a larger scale, the shape of the cliffs and the wave-cut platform at the base of the cliffs are worth studying. The different cliff profiles are due, in part at least, to the tilt of the rocks at the cliffline and hence the nature of the folding. Even the existence of some of the bays can be attributed to erosion of weaker parts of the limestone, for example West Angle Bay (SM 853033), where erosion has proceeded along the core of a downfold ('syncline'), leaving the more resistant limestone to form the headlands.

28   Mewslade Bay

Just to the west of the oil refineries, dockyards, steel works and mines of the Swansea region lies a small peninsula, Gower, which, criss-crossed by narrow roads and lanes, hosts some of the most beautiful coastal scenery in Britain. Gower is notable for its bays. The coast road from Swansea to the little resort suburb of Mumbles, once paralleled by a fine tramway, follows one of the largest of the bays. In fine weather the blue bay is dotted with sailing dinghies, and it exudes a Mediterranean air. Further west

another large bay is backed by an impressive set of sand dunes, while on the north side of the Peninsula, facing Llanelly across the Loughor Estuary, there are such bays as Broughton, with broad beaches, bays bounded by hills, and some wide salt marshes.

The less-visited, cliff-bounded bay at Mewslade is different again. To reach it take the little stony footpath, 0.75 km long, from Mewslade Cross near Middleton on the Rhossili road. The path descends along a small valley, shared in wet weather by a

trickling stream that disgorges onto the beach down a cleft in the grey Carboniferous Limestones. The valley sides are mantled by angular screes of limestone debris derived from the frost shattering of the scars above. Much of the vegetation is deformed and blasted by the salt-laden winds that blow from the south-west.

The west side of the bay is protected by a great headland formed in the massive limestone which, because of the inland dip of the rocks, provides a solid bastion against the attack of the

The fossil scree and raised beach platform at Mewslade

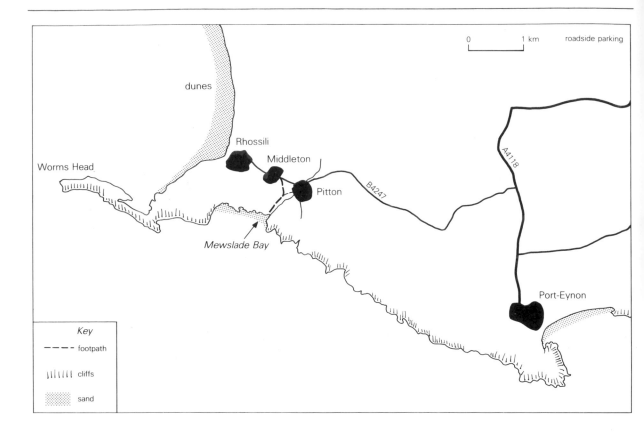

sea. It also serves as a safe home for nesting sea birds. The beach, which is almost wholly covered at high tide, nestles in the embayed, sheltered area between the headlands. In the rocks behind this beach there is a suite of features that merit a scramble.

At the lowest level, the scour of the tide, aided by the swirl of boulders, has created smooth pot-holes in the rocks. Above this level the limestones have been attacked by sea spray and by rain to give small pits, runnels, meanders and pinnacles. Moving up out of reach of the waves and tides, the rock is less pitted, and orange-coloured lichens flourish. Small plants, such as thrift, are able to colonise the rock surfaces, in which fossils embedded in the limestones can often be seen.

It is at this level that the tilting beds of limestone appear to have

been planed off, leaving an irregular platform well above the reach of the sea today. This is the first hint that the sea once stood at a higher level. Careful investigation of this platform confirms the impression, for resting on the surface are some clusters of rounded boulders and cobbles which were deposited by the sea. This raised beach occurs widely on the Gower coast, and was formed in one of the warm interglacial periods of the Ice Age when melting of the great polar ice caps caused sea level to rise 8 to 10 m above the present. It may well have formed around 120 000 years ago.

After the raised beach was formed, the climate deteriorated as another cold, glacial period started. Welsh and Irish Sea Ice caps entered the Gower Peninsula and other parts of South Wales. For some of this

time, Mewslade was right near the ice margin and suffered the full rigours of a periglacial climate. The cliffs were attacked by severe frost action and the masses of angular limestone scree accumulated on and above the raised beach platform. This scree was in turn cemented by lime-rich waters percolating through the joints in the limestone.

The view from the platform is spectacular. On a clear day the coastline of the South West Peninsula is framed by the massive headlands. One can listen to the surf, admire the wheeling sea birds, and scour the cliffs for small caves and clefts or delve into rock pools. There are no caravans and few people mar the scene.

29 Swallow holes at the Head of the Valleys

One of the great paradoxes of South Wales is the close juxtaposition of grim and declining industrial valleys with some fine unspoilt uplands. The original beauty of the valleys has been transformed by the iron workers and the coal miners, but the country at the heads of the valleys is some of the most attractive in Wales.

The broad bones of the topography can be traced in the geological cross section. The Brecon Beacons in the north are a mass of Old Red Sandstone (so called because they are the older of two major deposits of red sandstone in Britain) of Devonian age. Moving south, a succession of very different rocks are visible at the surface, becoming progressively younger. The coal mines that so changed the character of the area were developed in the Coal Measures, but between them and the Brecon Beacons two important rock types crop out: the Carboniferous Limestone and the Millstone Grit, both of Carboniferous age.

As in the Ingleborough area of northern England (site 7) or the Mendips near Bristol (site 46), the Carboniferous Limestone is associated with the development of some striking landforms, created by the solution of the limestone beds by rain and river. In particular, the area to the north of the coal-mining belt is densely pocked with depressions ('swallow holes') caused by the collapse of underlying caverns. However, what makes this rather different from most limestone areas is that the largest swallow holes occur not on the soluble limestones but on the largely insoluble Millstone Grit. Certainly, the limestone has many

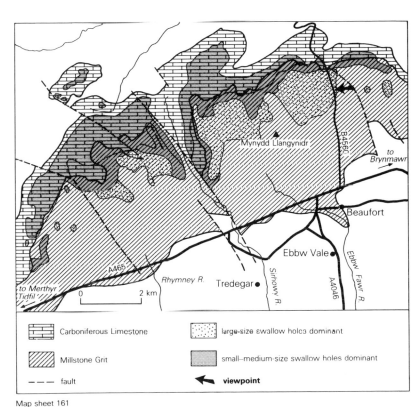

Carboniferous Limestone

Millstone Grit

--- fault

⬚ large-size swallow holes dominant

small–medium-size swallow holes dominant

⬅ viewpoint

Map sheet 161
Viewpoint grid reference: SO 158162

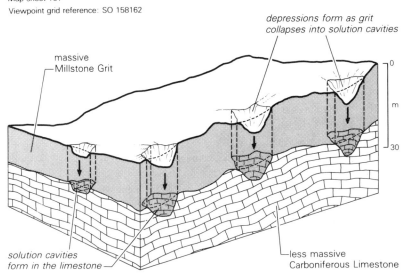

depressions form as grit collapses into solution cavities

massive Millstone Grit

solution cavities form in the limestone

less massive Carboniferous Limestone

The formation of large depressions by the collapse of Millstone Grit into solution cavities in the underlying Carboniferous Limestone in South Wales

hollows, but they are relatively small, being just a few metres deep and a few across. The hollows on the grit may be over 100 m in diameter and 20 m deep.

The explanation for this seemingly bizarre situation is to be found in the character and relative positions of the two rock types. The limestone is normally characterised by a closer spacing of joints than the Millstone Grit, so that beneath the grit the caverns that are formed by solution of the limestone can grow larger before collapsing. In other words, the grit, being less jointed, provides a more stable and extensive roof to the caverns that develop in the underlying limestone. Thus huge depressions are formed when the grit

eventually collapses into the cavern beneath. Caverns developing in the limestone where there was no cover of grit reached no great size before their well jointed roofs fell in. The resulting depressions or swallow holes are quite small.

Of the many and widely spread holes, the greatest concentrations of large holes occur to the north of the steel-making town of Ebbw Vale on Mynydd Llangynidr (SO 120140) and to the north of Penderyn near the village of Ystradfellte (SN 930135) at the head of the Neath Valley.

The latter locality is well worth visiting for another reason: the River Mellte flows underground here through a series of famous caves in the limestone. The river enters the

underground channel through the mouth of Porth-yr-Ogof cave, a seemingly innocuous entrance at the base of a bare wall of limestone. A short distance down stream it emerges from its treacherous route into a small, tree-lined gorge, where the gushing water has smoothed and polished its limestone channel, until in places it shines like glass.

Two paths lead down from the road, one to the entrance cave and the other in the opposite direction along the surface of the limestone to the exit gorge. Along the latter route pot-holes give glimpses of the foaming waters below, cascading through the limestone. The path also follows a small valley cut into the limestone surface and it is likely that

A large swallow hole, near the B4560 viewpoint

The Afon Mellte disappearing into Porth-yr-Ogof cave

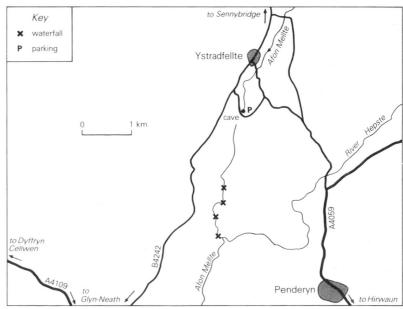

this was once a former, overground course of the River Mellte, possibly when underground flow was blocked by accumulations of ice (permafrost) in the limestones. Some of the rounding of the limestone along this path may also have resulted from solution at the boundary between the rock and overlying soil. This process of limestone dissolution has helped to fashion some of the classic limestone pavements, as for example in the Ingleborough area (site 8).

30 The shrinking Fens

The English Fenland is the only really large expanse of the British Isles that lies below sea level. It is a miniature Holland, and its present aspect results from the alternating fortunes of land and sea, fresh- and salt-water, man and nature.

About 5500 years ago the Fenland basin had relatively good drainage that was still above sea level (stage 1). Its surface consisted of Jurassic sandstones (Greensand) and clays (Gault), locally smothered in boulder clay and other products of the Ice Age glaciations. The flourishing vegetation cover included mature oak forest, and one fossil trunk from this period had a length of over 20 m without branching.

However, in Neolithic times the sea level began to rise and drainage conditions deteriorated. Peat (an accumulation of organic matter in fresh water) was formed in a boggy sedge fen (stage 2) at about 2700 BC, and this in turn was overlain by marine clay deposited from a shallow water, brackish (slightly salty) lagoon (stage 3). Towards the seaward margin of the Fens this clay is up to 7 m thick, but it thins out inland, where the salty water was shallower.

This incursion of the sea (called a 'transgression') did not last long because peat (stage 4) from *above* the marine clay (often called the Buttery Clay) is dated at 2200 BC inland and 2000 BC near the sea. However, the balance between the land and the sea shifted once more in Romano–British times (stage 5), when this peat in turn was overlain by marine silts. Unlike the first incursion, this one did not produce a large inland lake to be filled with brackish mud, but gave rise to extensive coastal salt marshes of silt, and to creeks with raised banks (called levees), which extended far inland into the peat country.

These ancient creek and river

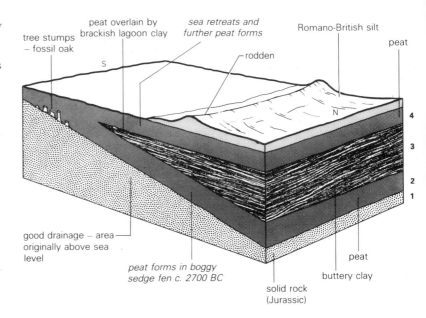

levees are marked to this day by '*roddens*' – sinuous silty lines or banks. These enable us to trace the former drainage lines. From time to time the remains of whales and similar large marine animals have been found in their silts, indicating that these ancient rivers were indeed estuarine. Lying above the general level of this flood-prone landscape, the roddens have been favoured sites for routeways and settlements. They are well seen from the air, but they can sometimes be picked out on the ground as lighter coloured, gently raised banks crossing fields of dark brown peat.

Although the Romans made many strides in the control of this area, it was from the 17th century onwards that the Fenland became more and more a man-made landscape. Artificial drainage ways were established, pumps were installed, and the area gradually became the rich and intensively cultivated agricultural area that we see today.

However, these 'improvements' have not been without consequence. One of these has been the gradual wastage and lowering of the peat deposits, for, as the natural water level in the ground (the 'water table') has been lowered by drainage, the peat has dried out and shrunk. This has encouraged both its chemical decay and the removal of its light, dry particles by the wind. Vast peaty dust clouds are often seen in early spring when the strong winds blow, unhampered by vegetation, across the open expanses of fenland. The soils are at their most vulnerable during dry spells before the new crops come through, and removal of trees and hedges, to aid large-scale mechanised farming, has added to the problem.

The rate of Fenland lowering can be witnessed by making a visit to Holme near Peterborough. Some time in 1851 or 1852, an iron post, from London's famous Crystal Palace, was driven through 7 m of peat into the underlying Buttery Clay. It acted as a sort of 'dip-stick' against which the degree of wastage could be gauged. In the following 80 years the peat shrunk by 3.3 m, though 2.5 m of this took place within 30 years of the drainage of nearby Whittlesea Mere. In these first 30 years after drainage and clearance of the natural vegetation, much of the surrounding area produced crops of rye-grass and even wheat. As the soil shrunk, however, arable farming became more and more difficult and most of the arable land had reverted to pasture by the 1880s. A depression in wheat prices at this time further encouraged the reversion. Trees such as oak, spruce and pine continued to be planted in shelter belts and game coverts, and by 1900 game coverts provided the main use. The birches were planted from 1902 onwards to give additional cover, but today they are the focus for the pleasant nature trail through the area that is now managed by the Nature Conservancy Council. A small mere is currently being recreated within the Reserve. The peat is shrinking very little today, and in 1983 the total amount of wastage was 4.25 mm.

In general, the level of the peatlands has fallen by about 5 m since the beginning of drainage and cultivation. Thus, the silt-covered Fens, which 17th century records show to have been about 1.5 m lower than the peat covered Fens, are today about 3 m higher than the drained peatlands.

Today the Fenlands are far removed from their natural state, except for a small patch at Wicken Sedge Fen (TL 550700) between Soham and Cambridge. In this delightful nature reserve the fen has been protected, and it is artificially maintained by pumping water in rather than out.

The old iron post from the Crystal Palace at Holme Fen

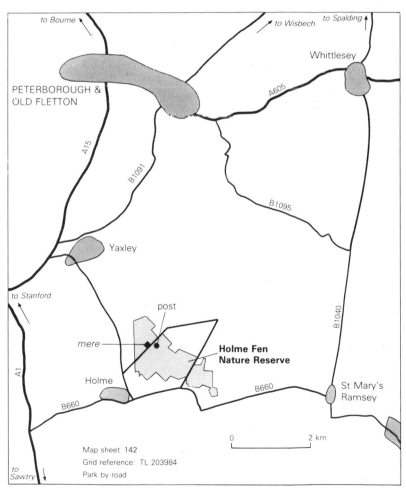

31   Walton Common's mysterious ramparts

During the Ice Age many of the parts of Britain that were not actually glaciated were, nonetheless, subjected to the rigours of an arctic or sub-arctic climate. Far removed from the moderating influence of the Atlantic, East Anglia was probably subjected to particularly severe conditions. Temperatures may well have been as much as 15°C cooler than today and the ground was frozen to great depths by permafrost.

Ice tends to segregate into layers and lenses as ground becomes frozen. This concentration of ground ice leads to localised upward heaving of the frozen ground surface above the ice, resulting in the development of a roughly symmetrical mound, called a 'pingo'. Such mounds may be quite steep; comparable features in Arctic

areas today, such as the Mackenzie Delta area of Canada and in East Greenland, range up to about 300 m in diameter and 60 m in height. In the summer months the ice in the surface layers of the soil melts, and the resulting soggy mass of soil and water flows down the side of the mound under gravity – a process called 'solifluction'. This and other mass-movement processes (such as creep; see page 8) will cause the material on the sides of the mounds to be moved downwards to collect at the base and in hollows between mounds. Thus, on later thawing of the actual ice lense, during a warmer phase, there will be relatively more soil material in the original hollows between mounds than on the sites of the former ice mounds. Hence,

hollows will result on the sites of former mounds and an inversion of relief will occur. The hollows formed on melting are termed ognips, and their presence indicates formerly intense permafrost

The formation of a pond and rampart by the growth and decay of an ice mound in the Ice Age

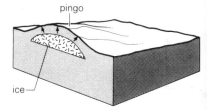

Ground heaved up over ice lens

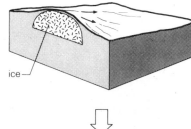

Soil is displaced sideways by solifluction

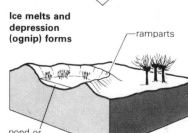

Ice melts and depression (ognip) forms

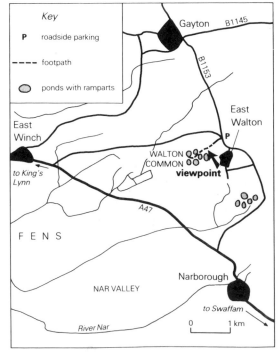

Key

P roadside parking

- - - - footpath

⬭ ponds with ramparts

Gayton   B1145

B1153

East Walton

East Winch

WALTON COMMON   **viewpoint**

to King's Lynn

A47

F E N S

NAR VALLEY

Narborough

to Swaffam

River Nar

0   1 km

Map sheet 132
Viewpoint grid reference:
TF 737165

An aerial view of the ramparts and hollows on Walton Common

conditions. They are in some respects similar to, and tend to be confused with, the hollows formed by solution in limestone.

Depressions and ramparts resulting from the growth and subsequent decay of ice mounds are found widely in East Anglia. Undoubtedly the best examples are to be seen at Walton Common (TF 735165), near King's Lynn in west Norfolk. The common is a low-lying area of rough grazing on the margin between the permeable chalk and the impermeable clay of the Nar Valley.

Many springs and water seepages occur at this boundary because the water draining through the chalk cannot sink into the clay and so it is forced to move sideways, along the boundary, finally seeping out on the edge of the hillside. This probably favoured ice mound development in the locality. The mounds seem to have formed successively in slightly different places so that the pattern is composed of mutually interfering ramparts and hollows. The latter, up to 100 m in diameter and 3 m in depth, often contain shallow ponds.

The freshness of these features suggests that they were probably formed at the very end of the last glacial episode, around 13000 years ago. Fortunately the uneven, swampy nature of the ground around them has saved them from the farmer's plough. There is a track leading from the B1153 down to the best examples on Walton Common; a well marked footpath crosses the common. Anyone keen on plants will find some interesting varieties in the boggy hollows.

32   Scolt Head Island

The western stretches of the Norfolk coast, seemingly rather flat and exposed to chill winds from the North Sea, have many devotees. They attract ornithologists who come to view migrant terns and waders; gourmets and gluttons who come to taste excellent sea food; yachtsmen who explore the winding creeks; botanists who value the rich flora of the marshes, and walkers who brave the mudflats to distance themselves from their fellow citizens. The most intriguing and yet little visited portion of this coast is Scolt Head Island, situated between Brancaster in the west and Burnham Overy Staithe in the east. At high tide it is an island, whereas at low tide it may be joined to the mainland, as Norton Creek, the main channel between Scolt and the mainland

marshes, sometimes dries out between House Hills and Burnham Harbour (see photo on p. 9).

The island developed as follows. First there was an extensive sandy foreshore with some shingle, similar in appearance to that exposed at low tide on other parts of the coastline today, notably near Holkham a few kilometres to the east. Continuous action of the waves, particularly during high seas, sorted the shingle from the sand, and in time piled it up into shingle ridges near the high water mark. The early ridges were unstable and mobile. Eventually one became more stable, dunes began to grow on it, and it gradually extended westwards, in several stages. Some of the shingle may have come by drift along the shore, but much has come from offshore, where there are ample

supplies in Ice Age glacial deposits. Each of former ends of the island are marked by curving lateral ridges of shingle, like the one seen today at the western end of the island. Some of these laterals, especially Long Hills and House Hills, carry high and well developed dunes.

The segments of the island between the laterals are occupied by marshes, and like the laterals the marshes change in age, and therefore in height, from east to west. The older marshes have had more time to trap silt and clay on their surfaces during the high tides, and so they are generally higher, and their creek patterns are better developed. However, in some cases improved drainage may have led to settling, and for this reason the oldest marshes may not *always* be the highest.

When walking over the marshes you may notice small basins called pans. These can result from creeks being dammed by the collapse of their banks. This then hinders the drainage, and the water in the pan will slowly evaporate – the residue becoming very salty. High salinity will inhibit vegetation growth, and so the floor of the pan remains bare. Other pans are not part of former creeks. They develop because vegetation spreads unevenly over a new marsh, often radiating from patches where seeds took root initially. In time a space may remain that has been completely surrounded by plants.

Scolt is a marvellous example of the way in which plants can actually help to fashion the landscape. Here the vegetation is varied and natural, unlike on many of the other marshes in southern England where the recently introduced species of cord grass (*Spartina anglica*) dominates the marsh. Once the bare marsh flat is formed, the first types of plants move in to colonise it – the simple patches of greenish algae. The algal masses trap silt from the sea, and seeds, and provide ideal conditions for the succulent salt-tasting marsh samphire (*Salicornia*), which then moves in. Eel grass (*Zostera*) is also an early coloniser, especially on sloppy muds. These plants encourage more deposition of mud by slowing down the mud-laden sea as it sweeps over them. Gradually, the clumps of vegetation become larger and the flow of tidal waters becomes constricted to certain channels – the creeks. The slightly increased height of the surface around the plants, due to their silt trapping abilities, leads to more favourable conditions, thus encouraging other plants to come in. These plants, such as sea aster (*Aster tripolia*), sea poa (*Glyceria maritima*) and sea blite (*Suaeda maritima*), are even more efficient at trapping and the height of the marsh continues to increase. New plants keep on appearing as the marsh grows, including sea lavender (*Limonium vulgare*), sea pink (*Armeria maritima*) and, along the creek margins, sea purslane (*Halimione portulacoides*). As the height increases, tidal inundation of the marsh becomes less frequent, the rate of growth slows down, and sea rush (*Juncus*) and black saltwort (*Glaux*) become the common types of plants. In all it takes about 200 years on Scolt Head to progress from the marsh samphire (*Salicornia*) stage to the sea rush (*Juncus*) stage.

Scolt may be reached by boat (the local fishermen run a service) from Overy Staithe (TF 845443) or Burnham Deepdale (TF 804443). You can also walk to it at low tide, from Burnham Deepdale across the salt marshes and Norton Creek. However, this is not recommended unless you have a good knowledge of the local tides and weather conditions, since it is easy to be stranded. Wellington boots are advisable when walking across the sticky muds of the marsh.

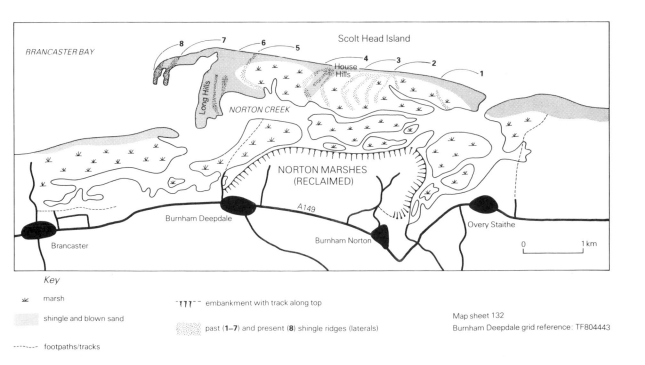

Key

⋎ marsh

shingle and blown sand

------- footpaths/tracks

⌐⊤⊤⌐ embankment with track along top

past (**1–7**) and present (**8**) shingle ridges (laterals)

Map sheet 132
Burnham Deepdale grid reference: TF804443

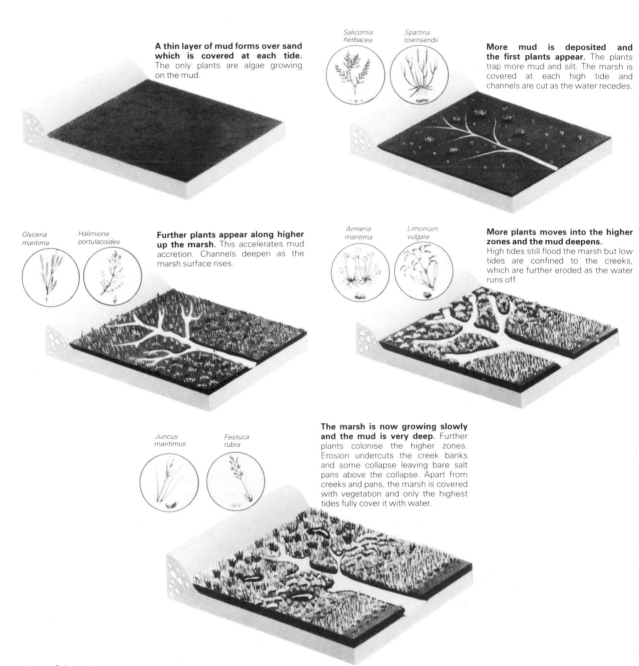

A thin layer of mud forms over sand which is covered at each tide. The only plants are algae growing on the mud.

Salicornia herbacea

Spartina townsendii

More mud is deposited and the first plants appear. The plants trap more mud and silt. The marsh is covered at each high tide and channels are cut as the water recedes.

Glyceria maritima

Halimione portulacoides

Further plants appear along higher up the marsh. This accelerates mud accretion. Channels deepen as the marsh surface rises.

Armeria maritima

Limonium vulgare

More plants moves into the higher zones and the mud deepens. High tides still flood the marsh but low tides are confined to the creeks, which are further eroded as the water runs off.

Juncus maritimus

Festuca rubra

The marsh is now growing slowly and the mud is very deep. Further plants colonise the higher zones. Erosion undercuts the creek banks and some collapse leaving bare salt pans above the collapse. Apart from creeks and pans, the marsh is covered with vegetation and only the highest tides fully cover it with water.

Stages of plant colonisation and marsh growth

33 North Norfolk: an Ice Age margin

Imagine yourself in Norfolk during one of the cold glacial phases of the Ice Age. The landscape is mysterious and frightening. Instead of the muddy waters of the North Sea, there is a great ice sheet, hundreds of metres thick, which from time to time creeps southwards into the interior of Norfolk. Beyond the ice front you shiver on the chill, barren, dusty plains, sparsely covered by scrubby shrubs, and crossed by streams that during the brief summer melt season become cobble-laden torrents. Picture the ice advancing and retreating several times, leaving behind it an inhospitable smear of infertile gravels and muds. Unsuited to agriculture, even today some of these deposits support little more than heathland wastes.

The great spreads of gravel dumped by debris-laden streams flowing out from the ice fronts (hence their name: outwash gravel) are particularly impressive. These are to be found at Kelling (TG 100420) and Salthouse (TG 070420), and are termed 'heaths' on the Ordnance Survey map. The Kelling Plain is the larger; the gravels can be traced past Holt (TG 080388) to Edgefield Heath (TG 083367), continuing on the other side of the upper Glaven Valley below Bunkers Hill (TG 086357), and then on to the area between Hunworth and Briston (TG 064334), and possibly in a degraded form to Briston Common (TG 062314).

The Salthouse outwash plain, to the west, is steeper and smaller, and it is bordered, in the Glaven valley, by a zone of curious hills. These look fairly ordinary, and one would never think that there was anything unusual in their formation until seeing a slice through them, revealing an abundance of loose gravels and sands, but no solid rock in sight. Muckleburgh Hill (TG 101429) and Great Hulver Hill (TG 064432) are

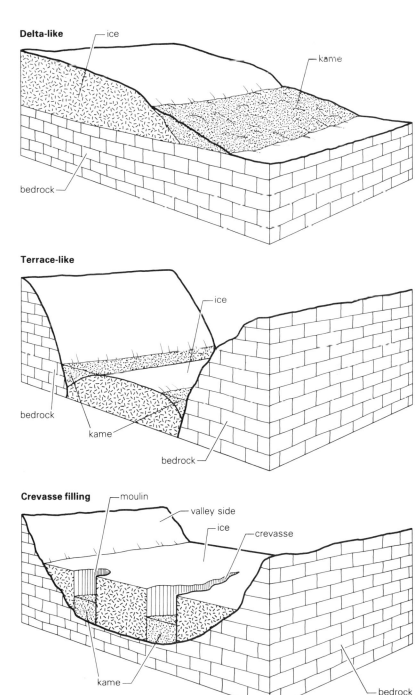

The cobble gravels of the Blakeney 'esker'

Map sheets 132 and 133

good examples. These are 'kames' and they form at the margins of – not beneath – a mass of stagnating ice. There are several types: some form when debris is dumped by streams at the edge of the ice front, building something like a small delta; where streams flow along the edges of the ice as it brushes against higher ground, debris may be dumped forming small terraces; elongated ridges of water-dumped debris may also form beneath crevasses in the ice front, and these are termed crevasse-filling kames.

An example of the last type is the so-called Blakeney 'esker', identified wrongly in the first place as an 'esker' (which is material dumped by rivers flowing *beneath* the ice sheet). This clear feature is about 3 km long and it extends in a northwesterly direction from the west side of the Glaven Valley. It runs parallel to other elongated kames, and exhibits sharp angular bends that reflect its origins in meltwater drainage through a system of crevasses in the stagnating ice. These meltwater drainage ways became choked with debris originally contained within and beneath the ice, the debris today preserving the outline of the former drainage routes. Some sections through the Blakeney 'esker' (called 'The Downs' on the 1:50000 Ordnance Survey map) can be seen on the Wiverton–Langham road at TG 031422. The effects of later frost churning on the cobbles within the poorly sorted debris are well seen in the sections.

The features described so far were mainly laid down by melt water. Other features in this scenically rich area of north Norfolk were dumped by the ice sheets themselves. Of these, the largest is the Cromer Ridge – a 100 m high belt of land extending from Cromer to Holt. Its northern slope is particularly steep, forming a very impressive feature in this area of gentle relief, especially near Sheringham (e.g. at TG 158420). The ridge, left at the margin of an ice sheet as it slowly melted and retreated, is a moraine, or a series of moraines. It is much larger than some of the moraines dumped by valley glaciers in upland areas (see site 18). The inside can be seen where it has been breached by the sea between Cromer and Weybourne (TG 110437). The waves are rapidly attacking these cliffs, which are largely composed of relatively unresistant sands and gravels, and the resulting shape of the cliff is rather different from that produced by erosion of the more clayey glacial deposits on the Yorkshire coast.

Considering that the features discussed above were formed in the penultimate ice advance or glaciation, they are very well preserved. Those in the area around Hunstanton are more recent, dating to the last ice advance, which affected only a small part of the county. The debris dumped beneath the ice sheet at this later stage – the boulder clay – is distinguished by its characteristic red-brown or grey-brown colour, and by the presence of erratics within it from the Cheviot Hills. During this episode, which was probably at its maximum around 18000 years ago, ice coming from the north and north-west locally overrode the coast and blocked a valley near Ringstead. This led to the formation of a small lake in front of the ice margin (a pro-glacial lake), since the outlet of the valley had been blocked. Eventually the lake contained sufficient water for it to overflow its lowest rim, which in this case was westwards. As the lake overflowed, the water excavated a gorge-like valley southwest of Ringstead – a valley which cuts across an older, broader valley, the course of which can still be traced. This 'overflow channel' is shown immediately north of Ringstead Downs on OS sheet 132. There is a track along the channel between TF 686399 and TF 707400.

A well defined esker in Hunstanton Park was deposited at roughly the same time by a glacial stream. Located at TF 695401, it is a continuous single winding ridge 2 km long, following a general north–south line. This ridge marks the former path of the meltwater stream beneath the ice sheet, the debris making up the ridge having been laid down in the bed of the stream. Access to this is difficult and permission is required from the Hunstanton Park estate office (at TF 688418).

34 The Norfolk Broads

There is probably no place in Britain more popular for the inland waterways enthusiast than the Norfolk Broads. Every year tens of thousands of the boating fraternity come to this group of otherwise peaceful lakes and ply from Broad to Broad. Why are the lakes there and how were they formed?

The broads lie in the valleys and tributaries of the Bure, Yare and Waveney rivers. These valleys contain long curves of alluvium and marshland, and find their exit to the sea through Breydon Water at Great Yarmouth. Over the centuries peat accumulated in these marshy areas. For many years it was believed that the broads were small natural lakes created by the uneven deposition of river sands and clays (alluvium) in these valleys, and that they were the remnants of formerly more extensive bodies of water that existed at times of higher sea level, during interglacials in the Ice Age. Indeed, this is probably the explanation for Breydon Water, which is a partially infilled estuary.

However, in the past few decades this traditional explanation has been discredited. It now appears that the broads are Britain's biggest man-

Barton Broad

made landform, and one of the most powerful manifestations of man's ability to change the landscape. Historical researches indicate that they result not from the action of men powered with modern earth-moving equipment, but from the patient digging of hundreds of medieval peat cutters, who excavated the marshland peats for fuel. The peat industry continued to be important until the end of the 19th century.

The evidence for this remarkable conclusion was pieced together from a wide range of sources. First, closely spaced boreholes revealed sudden lateral changes between the freshwater muds deposited today and vertical walls of peat or peat and clay. It is difficult to imagine a natural process that could account for this sharp break. Secondly, observations revealed that there were some curious upstanding islands or ridges of peat, which once again appeared to be bounded by vertical sides, and which had geometric forms and some clear alignments. The best explanation is that these represent balks of uncut peat left between adjacent diggings. Thirdly, documentary evidence of changes in names from turbary (a turf cutting area) to fen, to water, and then to marsh; the absence of early references to the broads in place names or in manuscripts; and the appearance in local records of valuable fisheries where previously there had been none, all indicate the artificial origin of the broads.

Their history is as follows. A marine invasion of the valleys during the Iron Age or Roman times led to the deposition of muds and clay over the existing peat in the valley floors. The clay deposition was widespread in the lower reaches of the valleys, but it petered out up stream. Peat continued to accumulate after the sea had retreated, covering the marine clay deposits and the earlier peats. When the peat came to be excavated, the best sites were those further upstream, where the

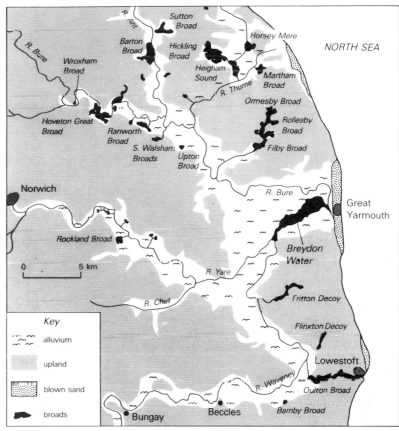

Map sheet 134

intervening clay layer was absent. The peat was cut to a depth of about 3 m, leaving basins which were later flooded, forming the broads.

The Broads are now under threat from the very agent that made them – man. Land drainage and developing agriculture, increasing populations in the river catchments and the need for disposal of their sewage effluents, and the development of the boating holiday industry, have all combined to change the character of the water bodies. In the past 20 years the water and water-side plants have begun to disappear at an alarming rate. The reasons for this transformation are

complex, and a combination of factors have probably contributed. Pollution from agricultural fertilisers seeping into the Broads, the discharge of fuel from the increasing number of diesel-powered cruisers, and the turbulence and resulting turbidity that their wash produces, have changed the environment in which the plants live. The loss of plants not only reduces the natural beauty of the area, but also removes a degree of protection from the banks of the Broads, and destroys the habitat of some of the region's bird life.

35   The Breckland Meres

In the heart of East Anglia, astride the borders of Norfolk and Suffolk, there is an area of sandy heathland and conifer forest called the Breckland. Its whole aspect is more like Luneburg Heath in Germany than lowland England. Within this distinctive landscape, with its windbreaks, its dark plantations and its flinty and sandy soils, there is a group of a dozen small lakes or meres. Lying in the catchment of the Little Ouse river, between Croxton and East Wretham, they are very beautiful and they attract a rich and varied bird life. But, for obscure reasons, the water levels in the meres fluctuate rapidly and frequently.

The largest mere is the kidney-shaped Fowl Mere (TL 880895), which usually occupies one major depression in the surrounding land, with a water surface of about 7 ha. At times the water overflows into satellite depressions to the east and west, occupying a maximum area of 12 ha. The mere is shallow, with a maximum depth of only 5 m, and its sides are gently sloping. Home Mere (TL 894897) is also shallow, but much smaller with an area of 1.8 ha. Unfortunately, both of these meres are located in a military training area.

The Devil's Punchbowl (TL 878892), which has public access and a good car park, is just 200 m south of Fowl Mere and is an approximately circular depression covering 0.6 ha. Compared with the other meres the Punchbowl is deep (about 7 m) and its slopes are steep (up to 18°). It is the most charming of all the meres, and an ideal location for a picnic.

Ring Mere (TL 910878) is in a nature reserve and, although the name might suggest otherwise, it is oval in plan, covering 2.4 ha just west of the Thetford–East Wretham road. 500 m to the north-west of Ring Mere and also within a nature reserve lies Lang Mere (TL 905885) which, when full, covers nearly 5 ha. It contains three separate depressions,

A late winter view of the Devil's Punchbowl

the largest kidney-shaped, the others oval. At times of high water level a small island exists between the depressions, and when the level is lower a ridge can be seen between the two smaller depressions.

The variation of water level is considerable, the meres frequently becoming dry, and less frequently overflowing. Long-term ranges in water level may be as much as 6 m, although annual ranges are usually 2–3 m. Surprisingly, levels are generally higher in the summer months from May to July, and lower between November and January. These seasonal fluctuations are, however, superimposed on longer trends, and since 1824 there have been thirteen periods, some of up to three years, during which most meres were dry, and eight periods, of up to four years, when water levels were very high.

The meres are underlain by chalk, which in this area may be covered by as much as 30 m of sandy material (boulder clay), most of which was deposited by glaciers in the Ice Age. Ground water occupies the available spaces within the chalk and the sandy boulder clay. The level of ground water below the surface varies according to the amount of rainfall, but, because it takes a period of time for rainfall to percolate into the ground and through the rock, there is often a time-lag between rainfall and the ground water response. This helps to explain why the water levels in the meres, which are fed by ground water, respond to high winter rainfalls some six to eight months later. Moreover, different meres respond in different ways, according to their size, depth, position with respect to the ground water level, and to how much boulder clay overlies the chalk. For

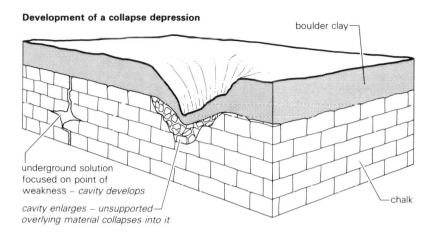

Development of a collapse depression

boulder clay

underground solution focused on point of weakness – *cavity develops*

cavity enlarges – unsupported overlying material collapses into it

chalk

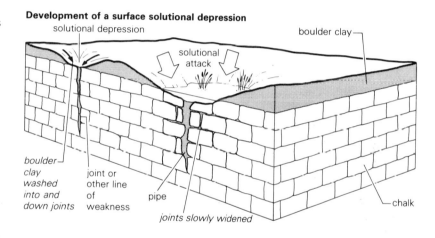

Development of a surface solutional depression

solutional depression

boulder clay

solutional attack

boulder clay washed into and down joints

joint or other line of weakness

pipe

joints slowly widened

chalk

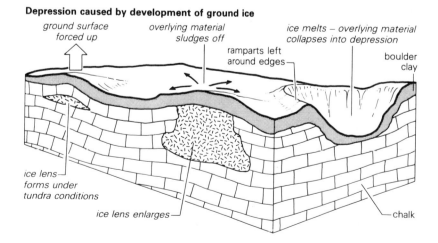

Depression caused by development of ground ice

ground surface forced up

overlying material sludges off

ice melts – overlying material collapses into depression

ramparts left around edges

boulder clay

ice lens forms under tundra conditions

ice lens enlarges

chalk

Some possible explanations for the formation of the Breckland Meres

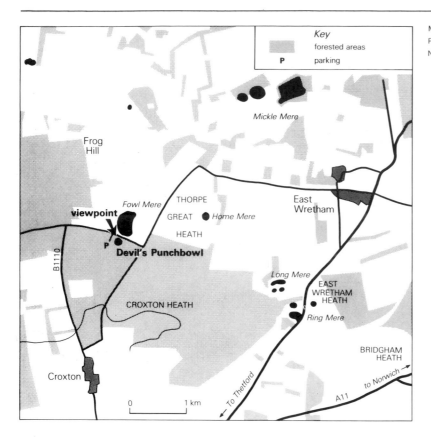

Map sheet 144

Punchbowl picnic area grid reference: TL 878893

Note MOD Danger Zone north of Thetford

these reasons one mere may be full when a neighbour is almost empty. The longer-term fluctuations are probably controlled by significant variations in the weather.

The origin of the meres has not yet been proven, nor are their ages certain. Three groups of ideas have been put forward. First, it has sometimes been thought that the meres were excavated by man. Pits resulting from extraction of limestone and marl for agricultural use are common in East Anglia, but they all seem to be smaller and more crater-like than the meres.

On the other hand, the depressions may have been formed by solution of the chalk. Chalk, like other types of limestone, is soluble in rain water and such solution is often concentrated at specific weaknesses, such as joints within the rock. The solution may be aided by the presence of ground water near the surface in this low-lying area and by organic acids derived from the pine trees and heath growing on the poor sandy soils. Solution certainly takes place, as shown by the small steep-sided holes developed in the limestone near the meres. The Devil's Punchbowl perhaps represents a transition between the larger, broad and shallow meres and the steep-sided solution hollows nearby.

There is, however, a third possibility that we need to consider, namely that the meres were formed by the disruption of the surface in the Ice Age by the freezing of water within the ground. Ice underground often concentrates into lenses and, since water expands on freezing, it exerts increased pressure on the ground and the surface is buckled, producing ice-supported mounds (pingos; see site 31). When the ice thaws, the mounds collapse, leaving a depression. The Breckland meres are rather larger than most pingo depressions, but in Siberia there are similar thaw depressions (called 'alases') of a size comparable to the meres.

Since chalk is highly soluble, and periglacial conditions have existed in the recent geological past, it is likely that both processes have affected the development of the meres.

36   More Breckland curiosities

On more than one occasion in the Ice Age, the great Scandinavian ice sheet crossed the present North Sea, linked up with ice caps radiating from the highland areas of Britain and spread out into the lowland areas. In East Anglia one of the earlier advances of this ice breached and eroded the chalk escarpment, producing a low-lying area of somewhat unusual drainage called the Breckland. The same ice sheets also deposited sandy and flinty debris or boulder clay, much of which was derived from erosion of the very sandy rocks near Sandringham in north Norfolk.

When later advances of the ice failed to reach so far south, the

Breckland area was subjected to extreme periglacial conditions, as described in site 31. Some of the traces of this frigid era can be seen in the landscape. For example, on Wangford Warren (TL 759841), at the far end of the great American airforce base of Lakenheath, there is a most surprising scene for inland England: an expanse of dune sand, rising some 6 m above the surrounding country, rather like a miniature desert. Most of the sand seems to have been blown out from the boulder clay to form dunes under the cold and relatively dry conditions during the last glacial phase about 18000 years ago. Inland dunefields

are rare in Britain; the only other examples are around Sandringham itself, in the Trent Valley north of Newark, and on the Wolds around Scunthorpe.

The orientation of the dunes is difficult to determine, as they are much dissected by 'blowouts' resulting when the vegetation cover degenerates for one reason or another. Thus, if the thin lichen crust over the soil, so well displayed at Wangford, becomes disturbed or broken, the loose dune sand is exposed to the further action of the wind. The vulnerable exposed sand is then scooped up and redistributed by the wind, leaving a great hollow in the original dune – a blowout. Much of the shape of the dune is lost as a result. A good blowout, still not recolonised by the stabilising vegetation and in which the orange–brown iron-rich layer of an ancient soil is exposed, can be seen at Wangford. Almost 1 km wide, this has its longest dimension oriented from south-west to north-east, reflecting the direction of the prevailing winds. Sand removed from the blowout is believed to have led to the overwhelming of Santon Downham in 1668; this village lies about 9 km north-east of the blowout and is directly in line with it.

Man's most recent response to the problem of wind action on the sandy soils in this area has been the planting of conifer windbreaks around farmland. These are especially common to the north-east of Thetford, in the parish of Kilverstone (TL 910850). However, much of the area is given over to larger plantations of conifers, these being an economic use of the poor quality soils. Before afforestation (beginning in about 1840), the sand was much more mobile than it is now, so that diarist John Evelyn (1677) compared the Breckland to the 'Desarts of Libya', and Gilpin wrote in 1769 of 'a piece of absolute desert almost in the heart of England'. More recently,

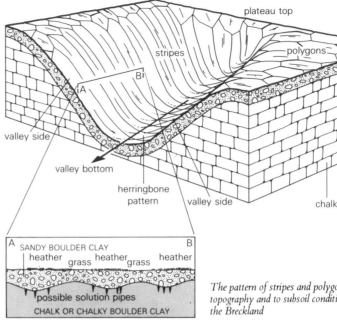

The pattern of stripes and polygons in relation to topography and to subsoil conditions as found in the Breckland

Clarke (1925) in his *Brecklands Wilds* wrote of 'a miniature Sahara' and put it down to the action of grazing rabbits destroying the vegetation.

A second example of the imprint of arctic conditions during the last glacial phase is provided by some bizarre stripes and polygons formed of alternations of grass and heather. The most appealing example is

where a herringbone effect has been produced by stripes on opposite slopes meeting and turning together along the floor of a shallow valley. These patterns are found on part of Thetford Heath and can easily be seen by taking the minor road from Barnham (TL 868792) towards Elveden (TL 822800). This road links the A134 and A11 south-west of Thetford. The patterns occur in fields to the north of the road around TL 848799.

Excellent and more accessible examples of more regular and simple patterns occur near those remarkable Neolithic flint mines at Grimes Graves, whence they are clearly visible. The best way to approach them is by driving a few hundred metres along the unmetalled track leading from the B1108 at TL 805901 and then walking across the field to the north of the road.

Heather and grass stripes near Grimes Graves

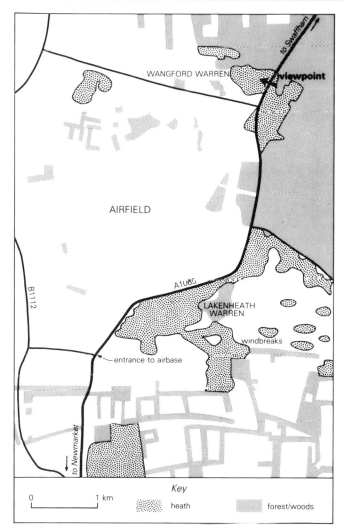

Map sheet 143   Viewpoint grid reference: TL 758841

Key
heath   forest/woods

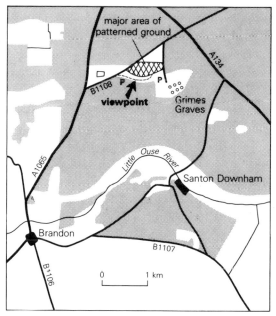

Map sheet 144   Grid reference of stripes: TL 809901

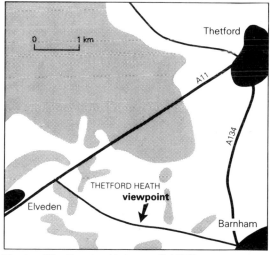

Map sheet 144   Viewpoint grid reference: TL 848799

Key
forest   P parking   ----- track

The locations of Wangford and Lakenheath Warrens (above), Grimes Graves (top right) and Thetford Heath (bottom right)

The exact formation of these patterns is not fully understood, though comparable features are forming today in Alaska by frost action operating on areas of tussocky birch heathland underlain by frozen ground (permafrost). It is thought that severe churning of the soil and subsoil occurs when the water in the ground freezes, for pressures are set up by the expansion of water on freezing. The effect of this churning on Thetford Chase led to localised concentrations of sand over the chalk, the concentrations having a regular polygonal pattern. This pattern is reflected by the vegetation; the heather, which likes acidic soils, occurs over the sand, whereas grass tends to prefer places where the chalk is uncovered or the sand cover very thin. The polygons characterise flatter areas and are elongated on steeper slopes into stripes.

101

37   The Devil's Kneadingtrough

The village of Brook, with its delightful church, oasthouses and wooden barns, lies close to the edge of the chalk escarpment of the North Downs. The escarpment here has been cut into by a cluster of seven coombes or dry valleys; striking, steep features that run back up to 450 m into the escarpment. They are from north to south; the New Barn Coombe, The Old Limekiln Coombe, Fishpond Bottom, Newgate Scrubs, the Devil's Kneadingtrough, and the two small Piclersdane Scrubs Coombes.

Of these, the Devil's Kneadingtrough (TR 076453), a nature reserve managed by the Nature Conservancy Council, is accessible along a thorny and often rather muddy track that starts 25 m or so west of the 'T' junction just north of Brook. Following the winding track up the western side of the valley there are excellent views of the trough's box-like bottom and its steep sides clad in tussocky grass. Beyond the scarp stretch the patchwork fields of the vale, into which the valley runs through an area of disturbed and hummocky ground. A good viewpoint for the less energetic is easily reached from the parking spaces on the road to Hastingleigh. Follow the footpath over the stile and the Devil's Kneadingtrough suddenly falls away in front of you.

The seven coombes all have flat or very gently concave floors in cross-section, and steep, rather straight valley sides which vary in angle between 20° and 34°. They have down-valley gradients approaching 17° at their upper ends, grading down to 4–5° in their lower parts. Channels, occupied by thin trails of chalky debris, cascade down the steep back walls. The trails widen down slope to form broad tongues of rubble which occupy the floors of the coombes, and extend as fans on to the vale beyond.

Excavations into and beneath this

The Devil's Kneadingtrough cutting into the Chalk escarpment of the North Downs

chalky debris have enabled scientists to establish the history of the coombes. The sequence of events was as follows. Towards the end of the Ice Age the climate gradually became milder, and arctic conditions receded. Frost action became less fierce, and thin soils and organic layers accumulated on the surface of the chalk. There was then a brief reversal to a harsher, colder climate, and these organic deposits were overwhelmed by a great fan of chalky debris, supplied by the action of frost and ice breaking up the chalk. Subsequently, the warm climates of postglacial times returned and the slopes became stable and were mantled with soil and forested. Radiocarbon dating of the chalky fan debris suggests that it accumulated in a mere 500 years (8800–8300 BC).

A study of the volumes of fan debris of this age, and the volumes of the coombes themselves, suggest that as much as one third of the chalk which must have been eroded to form the valleys is still present within a radius of less than 1.6 km of the scarp. Allowing for the fact that quite large amounts of chalky material might have been removed beyond this distance, and that much may have been destroyed by solution, then the great bulk of the erosion of the dry valley cluster must have taken place during this short interval.

Study of fossils and the work on deposits of similar age elsewhere in Europe, suggest the climate at this time was both cold and humid. These conditions favoured freeze–thaw activity and solifluction. It is possible that, in full glacial times, conditions were insufficiently humid for these processes to have the same effect. The freeze–thaw acts to prise blocks of chalk away from the valley sides, and the solifluction transports the blocks down slope in a slurry of debris during summer melting of the ice. The solifluction therefore gives rise to the fans of debris that make up the hummocky ground on the edge of the vale.

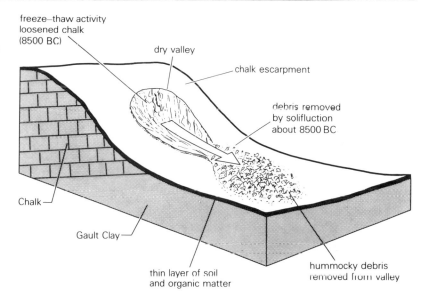

The evolution of the Devil's Kneadingtrough

Map sheet 179
Viewpoint grid reference: TR 076453
Roadside parking

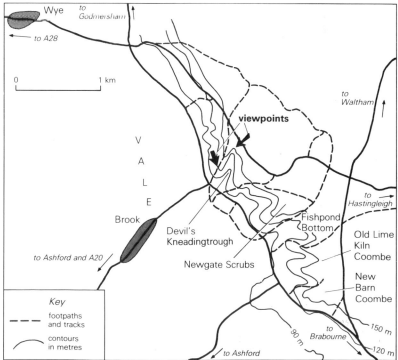

38 Folkestone Warren: the railway's lost property

The white chalk cliffs of the south coast of England are often seen as imposing and impregnable barriers that help the Channel to keep our shores free from foreign invasion. However, in some places the chalk cliffs seem far from stable, and prone to rapid and calamitous collapse into the sea. One such locality is an area of land lying just to the east of Folkestone: Folkestone Warren (TR 250380). Here we see not a simple bold cliff facing out across the Channel but a much more jumbled up piece of scenery on whose seaward edge man has plainly been forced to build extensive and ugly sea walls and other defences against the waves. Through the middle of the area there is a railway, a line of chequered history.

The cliffline has been subjected to a series of major landslides. The prime reason for this is that the nearly horizontal beds of chalk here overlie a particularly thick (around 45 m) bed of unstable Gault Clay. Slides of masses of chalk over the Gault Clay are known to have occurred in 1716 and 1765. They also occurred sporadically in the 19th century, but the greatest and most dangerous foundering of the cliff took place in 1915. On 19 December of that year a great slide, which travelled more than 400 m out to sea, buried the railway line under about 13 m of debris for a length of about 250 m. Great rafts of chalk fell away from the main cliff, and one house, called Eagle's Nest, slid 30 m gently downwards on one such raft. The movement was so gentle that the house was scarcely damaged, but in front of the main slips the sea bed was upheaved, creating ridges up to 350 m from the former shoreline.

Both the road and rail communications were disrupted. The old main

Disruption of the railway line through Folkestone Warren in 1915

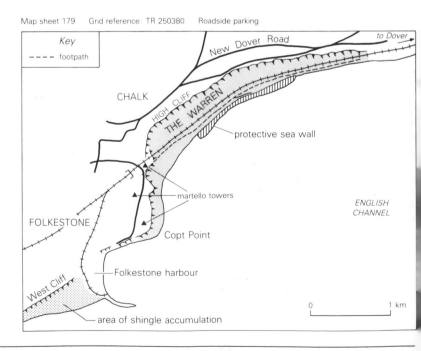

Map sheet 179   Grid reference: TR 250380   Roadside parking

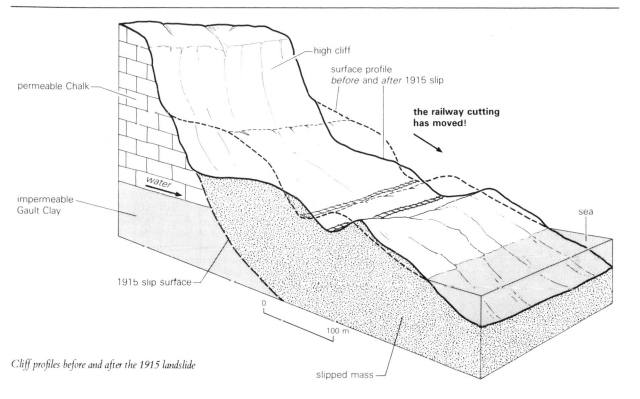

permeable Chalk

high cliff

surface profile
before and *after* 1915 slip

**the railway cutting
has moved!**

impermeable
Gault Clay

water

sea

1915 slip surface

0

100 m

Cliff profiles before and after the 1915 landslide

slipped mass

road from Folkestone to Dover was diverted, and a train from Ashford to Dover, carrying about 130 passengers, was mercifully late and was stopped by soldiers waving red lights. Nonetheless, as the passengers sat in the stationary train the engine and front carriages gradually sank down as the slide affected the line.

Further movements have occurred since 1915, but between 1948 and 1955 a massive wall was constructed in an attempt to halt them, and it appears to have been largely successful. However, even though man may now be able to control this slide, he was also in part responsible for its past activity. The construction of Folkestone Harbour, particularly

between 1810 and 1905, changed the pattern of erosion and deposition along the coastline. The harbour works trapped sediment that was being moved by the waves along the coast from the west. Consequently, a shingle deposit built up on the westward side of the Folkestone breakwater at West Cliff, but the beaches were robbed of shingle down drift (i.e. eastwards) from Folkestone Harbour. This in turn exposed the Gault Clay in the cliffs at the Warren to greater wave attack, which in due course made it – and the whole cliffline – unstable. An account of *Folkestone and its neighbourhood* by S. J. Mackie in 1883 records the situation well:

'Beautiful as the Warren still is, it is not to be compared with the Warren of half a century ago. The ingenuity of our present generation, which has constructed a harbour where our forefathers utterly failed, has not been without effects which lovers of the grand and beautiful scenes of nature must regret. While the beach has accumulated to an enormous extent west of the harbour, the rugged form of Copt Point has crumbled away until nothing is left but the rocky foundation on which it stood, and year by year the Warren which nestled behind it grows less and less in extent, as the restless waves roll up and swallow it.'

39 The Seven Sisters and Beachy Head

On the Sussex coast, between Eastbourne in the east and Cuckmere Haven in the west, there are some of the most impressive chalk cliffs in the whole of southern England. Whereas at Folkestone Warren in Kent (site 38) the cliff line has been worn down and complicated by major landslipping of the chalk over the underlying Gault Clay, in this area the chalk cliffs rise up majestically.

The chalk is a form of limestone that was laid down in horizontal layers (strata) beneath the sea 100 million years ago. As a result of gentle folding about 25 million years ago (when the Alps were formed), these layers now dip at a slight angle. At Seven Sisters the cliff line is developed at right angles to the dip. The remarkable feature of these

cliffs, however, is that they have a distinctive switchback crest with seven rounded summits – the Sisters. This has made them an unmistakable landmark for generations of seafarers.

This switchback form originated during the Ice Age when a river system became established on the gently dipping chalk. Like so many other valleys in southern England (e.g. site 44) they are now dry, for the water drains into the permeable chalk. Under periglacial conditions in the Ice Age, however, the joints in the chalk would have been sealed by ice, and so the water would have flowed in rivers across the surface of the rock. The rivers probably flowed to a sea level lower than now, for at the time much of the world's water

was trapped in vast polar ice caps. When sea level started to rise during the last 12 000 or so years, it drowned the Ice Age river valleys, forming an indented coastline, as seen in parts of Cornwall (site 64) and South Wales (site 26). However, in this case the rocks are less resistant and the indented coastline was gradually straightened by waves first attacking the vulnerable headlands and then the whole cliffline. As the coast retreated, so the dry valleys were left hanging, forming the low points on the crest of the cliff line.

The dry valleys are only short. The longest, Gap Bottom, is 1.8 km long, and at Crowlink it hangs only 18 m above the beach. The shortest valley, appropriately called Short Bottom, is only 400 m long. Birling

A view of the Seven Sisters from the shore platform near South Hill

Gap, at the mouth of the Wish Valley, marks the eastern end of the Seven Sisters; it occurs at a lower level than any of the dry valleys separating the Sisters.

The great whiteness of the Seven Sisters, their vertical faces, and the general lack of rubble or debris beneath them, reflect the work of the sea. The cliffs are retreating at between 0.5 and 1.25 m per year, though much of the retreat takes place in occasional catastrophic collapses. In 1925, for example, half a million tons of chalk dropped onto the foreshore at Baily's Brow producing a minor earth tremor! The erosion and collapse is facilitated in this area because the chalk dips slightly towards the sea. Thus marine erosion at the base of the cliffs renders unstable much of the cliff face above. As the cliffs retreat, they leave behind a planed-off platform of chalk – the shore platform – at their base.

The imposing cliffs at Beachy Head are rather different in form than those of the Seven Sisters. Between points TV 582953 and TV 594955 they rise to 160 m and cut right through the South Downs escarpment. They are the highest cliffs in south-east England, and they provide a backdrop for the well known lighthouse on the shore platform. It is important to remember that these cliffs are not only high but also very dangerous. People die at Beachy Head every year through sheer carelessness. One should always keep at least 3 m back from the edge of the cliffs.

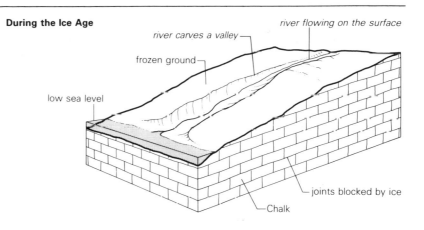

During the Ice Age

river flowing on the surface

river carves a valley

frozen ground

low sea level

joints blocked by ice

Chalk

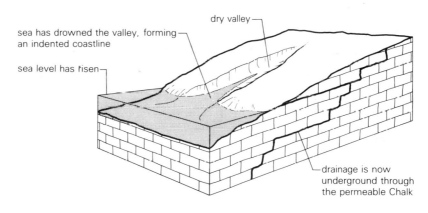

About 6000 years ago

dry valley

sea has drowned the valley, forming an indented coastline

sea level has risen

drainage is now underground through the permeable Chalk

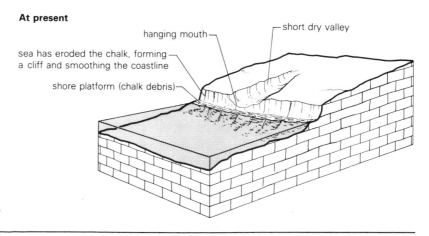

At present

hanging mouth

short dry valley

sea has eroded the chalk, forming a cliff and smoothing the coastline

shore platform (chalk debris)

Stages in the evolution of part of the Seven Sisters coastline

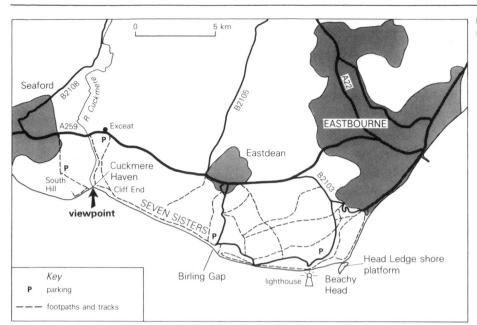

Map sheet 199
Birling Gap grid reference: TV 554960

Access to the area is relatively easy. The Seven Sisters may be approached from either west or east. There are car parks at both Exceat Barn (TV 518995) and at Birling Gap (TV 554960) which provide access to a path along the cliff. Good views of the Sisters across Cuckmere Haven can be obtained from South Hill (507978). To reach the cliff top at Beachy Head park on the nearby road and scramble down the rough tracks near the coastguard station (TV 591958) on to the cliff top footpath. One km NE along the footpath is Cow Gap (TV 596957) where there is access to the cliff foot. The footpath continues to Eastbourne. Head Ledge, which offers a panoramic view of the Beachy Head cliffline, is accessible at low tide across the shore platform.

40   The High Rocks of Tunbridge Wells

Just outside the fashionable spa of Tunbridge Wells, in a woodland setting on the south side of a little valley of a tributary of the Medway, there is a range of tall sandstone cliffs or tors split up by remarkable chasms – High Rocks (TQ 559384). These rocks became a popular excursion from the spa after they had been visited in 1670 by the Duke of York, who later became James II. Today they are visited not only by those who enjoy the varied autumn tints of the oaks, pines, yews, holly and rhododendrons growing among the rocks, but also by mountaineers who come to practise their rock-climbing skills on the 10 m high, lichen-covered cliffs. Access is through a lych-gate of romantic design opposite the High Rocks Hotel. Within, rustic walkways and bridges lead enchantingly from viewpoint to viewpoint.

It is the beautiful setting of High Rocks that makes these inland cliffs so much more attractive than equally important examples elsewhere in the Weald. Such features are commonly found wherever the upper bed of the Lower Tunbridge Wells Sandstone, the Ardingly Sandstone, occurs. Also worth a visit though is Chiddingly Wood (TQ 347322), with its perched block called Great-upon-Little (TQ 348322).

How were these cliffs made? The main factor is the nature of the Ardingly Sandstone, a sugary, rather soft sandstone cut by numerous large joints. These joints have widened to produce features that quarrymen call gulls, which when opened up by erosion can leave a steep-sided labyrinthine pattern of isolated blocks. It appears that downcutting of the local stream caused the valley slopes to become unstable so that blocks of sandstone tended to slip towards the valley sides, leading to the opening of the gulls – a process

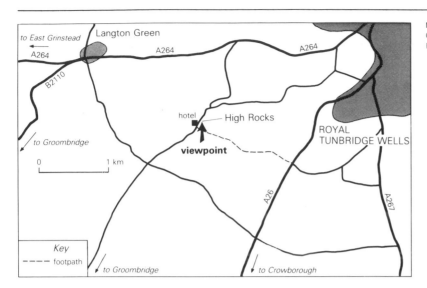

Map sheet 188
Grid reference: TQ 559384
Roadside parking at hotel

called 'cambering'. The growth of large trees, such as yews, with strong penetrating roots, may have helped to prise the blocks apart and widen the joints.

The shape of the cliffs also partly results from the undercutting of the blocks of sandstone. The cause of the undercutting has been the subject of long debate. Some early scientists felt that the sea was responsible, whereas others favoured man's intervention or the blasting effect of wind and sand under desert conditions. It is quite possible, however, that the undercutting results from faster erosion of the sandstone at the base of the cliffs. This may be caused by seepage of ground water or the presence of a less resistant silty band within the sandstone.

An interesting characteristic of many of the sandrock features is the development of small patterns on their surface by the action of weathering. One such feature is a hard brown rind, generally a few millimetres thick, which gives a resistant shell to the outer face of the rock, but weakens the inside. This is probably the result of mineral salts being drawn by evaporation from the interior of the rock to the outer skin. Another minor weathering form is the system of rectangular and hexagonal cracks reminiscent of a tortoise shell. These occur on the upper edges of boulders and rock outcrops, but their origin is uncertain. More frequently one finds various cavernous forms, including honeycombs and small pits, often restricted to the sides and overhanging section of the rocks.

The age of the sandrock features is puzzling. Some people believe they formed during arctic or periglacial conditions in the last Ice Age, and suggest that at present they are being covered up. Indeed, examples are known where sand, apparently produced by contemporary weathering, is banked up at the base of the cliff. However, the evidence is conflicting, for in other places angular debris of presumed periglacial origin is banked against them. Elsewhere, excavation of the buried cliff is occurring at the present day, and gullies are commonly found cutting into and removing the angular debris. The origin and age of these cliffs is therefore still open to question.

41   Symonds Yat and the meandering Wye

At Tintern Abbey and other beauty spots on the River Wye between Chepstow and Ross, the visitor is struck by the remarkable way in which the river has cut down deeply into the surrounding land. The incision itself is notable, but still more intriguing is the way in which the river almost turns back on itself in a series of giant meander loops. Some loops cross not only from lowland to highland but over different rock types too.

Some time in the past, sea level was perhaps 150 m above its present level, and the river developed a broad valley and flood plain, up to 1.6 km wide, at this high level. The mature river meandered across this flood plain, as most rivers do, with the meanders gradually migrating downstream. Migration of the meander bends takes place because the major currents flow around the outside of the bends, along the deeper parts of the river channel, being especially powerful just beyond the apex of the meander curve. Where these flows scour the outer river banks river cliffs are formed, and continued erosion and undercutting of these leads to their collapse. In this way the river shifts its channel slightly sideways and downstream, and the meander bends slowly migrate. Deposition of sediments takes place in the quieter waters on the inside of the meander bends, creating gently sloping areas.

The serenity of the meandering Wye was dramatically upset when the whole area was uplifted relative to sea level. The resulting increase in gradient gave the river renewed erosional energy – a mature river was rejuvenated. The Wye adjusted to this by starting to erode downwards into its valley floor,

The River Wye from the viewpoint at Symonds Yat

111

River is 'rejuvenated' as sea level falls, and begins to deepen its valley within an earlier floodplain

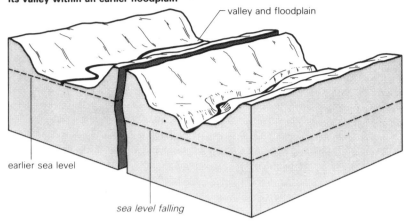

valley and floodplain

earlier sea level

sea level falling

Valley evolution as a result of river rejuvenation, meander incision and drainage superimposition

Meanders migrate down stream

floodplain remnant
river cliff

steep cliffs being cut by river flow

ingrown
meander

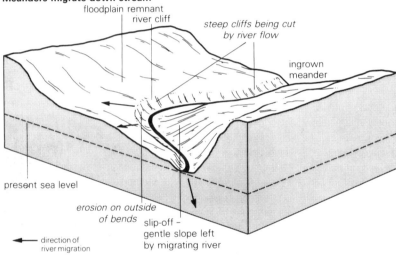

present sea level

erosion on outside of bends

slip-off –
gentle slope left
by migrating river

← direction of
river migration

Rapid downcutting produces an entrenched meander

entrenched meander

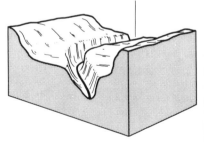

A layer of newer rock is eroded by a river, which maintains its course as it cuts down into older rock

newer rock

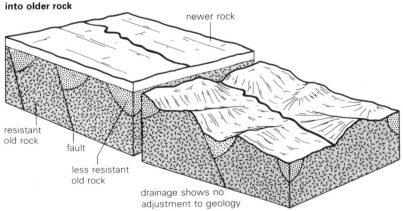

resistant
old rock

fault

less resistant
old rock

drainage shows no
adjustment to geology

producing a 'valley within valley' form that is so characteristic of rejuvenated landscapes. The meanders became carved into the landscape. However, the downward incision took place relatively slowly, probably because the rocks were fairly resistant and the uplift slow. The river therefore continued to erode sideways as well as downwards, with the result that the bends migrated downstream and were simultaneously deepened by downward erosion. These forces of erosion carved huge river cliffs around the outside of the bends, while deposition continued on the inner edges of the meanders. Thus, the interaction of these complex 3-D forces of erosion and deposition has created highly irregular and asymmetrical cross profiles in the valley.

Meanders formed in this way are described as 'ingrown', and good examples can be seen at St Arvans (SO 538965) and at Symonds Yat. Their shape contrasts with the sinuous, relatively narrow trenches formed when erosion is predominantly downwards. This can happen when the rocks are weak and the uplift rapid. The 'entrenched' meanders so formed are relatively symmetrical in cross profile.

Besides being rejuvenated, the Wye also provides an example of 'superimposed drainage'. The course of the river and its tributaries show virtually no adjustment to the underlying geology or to relief, as observed today. For example at Symonds Yat the Wye flows through the hilly area on to the lowlands and then back again into the hills, all in the space on one ingrown meander loop. Moreover, it crosses from the Carboniferous Limestone rocks of the higher land to the Old Red Sandstone and back. Thus the relief of the area partly reflects the rock types, but the river pays no heed to either. Where such a situation exists it is reasonable to suspect that the river evolved on a cover of more recent rocks which have

subsequently been removed by erosion, exposing older rocks with a quite different structural pattern. The river course is, therefore, slowly superimposed on the older rocks.

Symonds Yat (SO 564160) is the most famous viewpoint in the Wye Valley. It takes its name from Robert Symonds, a former high sheriff of Herefordshire, who owned extensive property in the area in the 19th century, and from a local word, 'yat', meaning a gate or a pass. A well worn path leads from the car park to a high vantage point in the narrow neck of the meander loop. All of the

features mentioned above can be seen from here, in addition to a great expanse of landscape beyond. The valley itself is very beautiful, and a walk down to the village of Symonds Yat is well worth the effort. The drive from Ross to Chepstow, mostly along the Wye Valley, is also recommended.

Map sheet 162

Viewpoint grid reference: SO 564160

Parking, car park (**P**) in cleared land just off B4432, opposite Yat Rock Café at Yat Rock

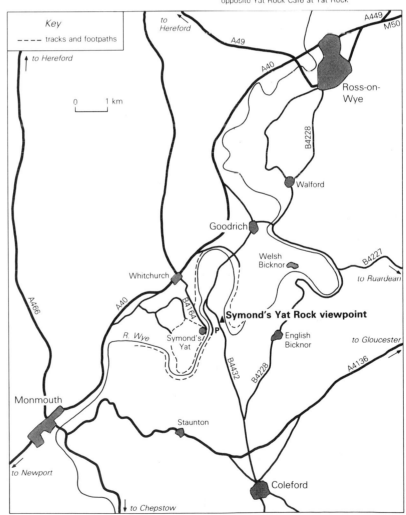

42   The slipping sides of Cleeve Hill

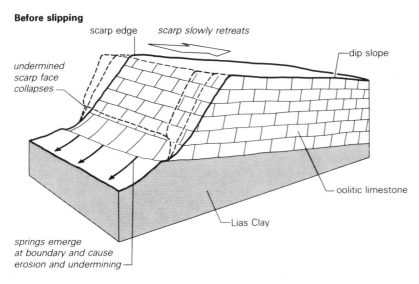

Before slipping

scarp edge

scarp slowly retreats

undermined
scarp face
collapses

dip slope

oolitic limestone

Lias Clay

springs emerge
at boundary and cause
erosion and undermining

The formation of the slip troughs on the sides of Cleeve Hill

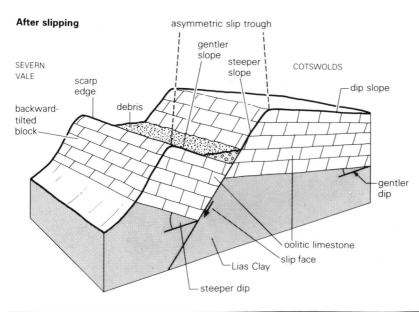

After slipping

asymmetric slip trough

gentler
slope

steeper
slope

SEVERN
VALE

COTSWOLDS

scarp
edge

debris

dip slope

backward-
tilted
block

gentler
dip

oolitic limestone

slip face

Lias Clay

steeper dip

To Cleeve Hill, the highest point of the Cotswolds, the inhabitants of Cheltenham and its surrounding villages come in their hundreds by foot, by car and by bus, partly to escape the sultry humidity of a summer's afternoon in the vale – the same weather that attracted the fabled peppery Anglo-Indian military men to retire there – and partly to appreciate the notable views. On a clear day one can see some tens of kilometres across the vale of the Severn to the older rocks of the Forest of Dean, May Hill, and even to the Sugar Loaf and the Black Mountains near Abergavenny in Wales. The Malvern ridge rises above the plain like a miniature Alps, while to the north the smoky plateau on which Birmingham lies can sometimes be seen. Here one can encompass both highland and lowland Britain in one broad sweep. One can also appreciate some of the wildest and most bracing walks across the short sheep-cropped turf of the Cotswolds, and investigate some of the many abandoned quarries that expose the golden Cotswold limestone.

The Cotswold Hills are one of the main scarps of lowland England, and like the others, such as the Chilterns, they are formed where relatively resistant sedimentary rocks rise up above vales developed in weaker clays. They are composed of a limestone which is usually termed 'oolitic', from the Greek word for egg: the texture of the rock reminded early observers of the roe of a fish. It is stronger than the chalk, but generally less resistant than the Carboniferous Limestone that makes up the more classic limestone country such as the Mendips (site 47) or the Malham area (site 8). Unlike the latter it contains very few underground caverns, and even the

Slip troughs developed in the main Cotswold escarpment near Lower Hill Farm

few small caves along the scarp are little more than slightly widened joints. However, as in many limestones, much of the drainage is underground.

The Cotswold limestone, which is around 150 million years old, overlies the impermeable and much less stable clays of Jurassic age. Along the escarpment, at the boundary of the clay and limestone, springs are thrown out: whereas the limestone contains many joints, fissures and pores which can hold and transmit water, the clay is devoid of such passageways, so that water percolating down through the limestone cannot sink any further. Thus it emerges at the boundary between the two highly contrasting rock types. As the water flows out it causes erosion and undermining of

the rocks, which may then collapse. In this way the scarp retreats, and the springs play a role in creating great embayments, such as those in which Cheltenham and Winchcombe lie.

Because the sapping action of the springs has formed such a steep escarpment, and because the limestone rests unstably on the rather mobile blue-grey Lias clays, Cleeve Hill has been the location of a large number of rock slides. Some Victorian geologists were so impressed by these slides that they proposed that the scarp was formerly a sea cliff that had been attacked by waves when the Earth was inundated by Noah's Flood. The fact that the rock layers have been tilted and disturbed by slipping (cambered) can be seen in many of the quarry sections along the scarp face (notably

at SO 987272). Rather more impressive expressions of the slipping are provided by groups of troughs sunk into the landscape; one group occurs along the main scarp face above Prestbury and Southam, and another group truncates several spurs near Postlip Warren.

The troughs above Prestbury and Southam are best developed at two localities: to the south-east of Lower Hill Farm (SO 992237), and to the west of the radio masts (SO 992248). Both troughs run approximately parallel to the main trend of the escarpment. The trough near Lower Hill Farm is approximately 500 m long, starting as a small depression at the scarp face. It then runs slightly away from the edge and becomes entrenched to a depth of about 14 m before heading back to the escarpment face and ending in a large, steep embayment. It has a slightly irregular long profile, with ups and downs, and its sides show marked asymmetry, the plateau side being steeper. The trough to the west of the radio masts is much smaller and straighter and less like a canyon. However, it too has a very asymmetrical cross profile.

The Postlip Warren group of troughs at SO 997265 consist of three striking features truncating the spur between the Postlip and Corndean valleys. They are approximately parallel, are 12–15 m deep, and are aligned at right angles to the major dry valleys dissecting the plateau, but parallel to the great embayment in which Winchcombe lies. The bottoms of these troughs are broad and flat (50–70 m wide) and, once again, they have asymmetrical cross profiles.

These curious features have certain points in common. They are all asymmetrical, have irregular or closed long profiles, run parallel to the main relief trends, truncate major drainage lines, and in some cases there is little or no area draining into them ('catchment area'). The most satisfactory explanation is that the troughs are produced by the foundering of massive, joint-controlled rafts of oolitic limestone on an unstable base of moving clay. In the process many of the rafts or blocks of limestone were tilted backwards – termed a rotational slip (as in site 55). The faces along which the slips took place are relatively steep, and this explains the asymmetry of the troughs. One would also expect the troughs to develop parallel to the edges of either the escarpment or the great embayments within it and, in the case of multiple slips, for the troughs to be parallel to each other. As they are not valleys formed by rivers, it is not surprising that they have little or no catchment area and that they cut across the main dry valleys of the area.

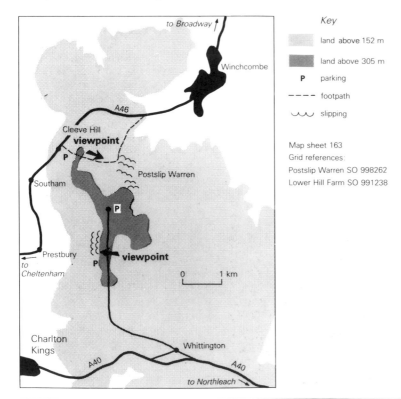

Key

land above 152 m

land above 305 m

P   parking

– – – –   footpath

‿‿‿   slipping

Map sheet 163
Grid references:
Postslip Warren SO 998262
Lower Hill Farm SO 991238

43 Cotswold misfits and the leaking Leach

Landscape tastes tend to vary over time. In the early 19th century, during the course of his *Rural rides*, William Cobbett dismissed the elegant spa town of Cheltenham that had grown up during the Regency as an abomination inhabited by some of the least salubrious types of crook and *nouveau riche* that one could find anywhere. He was scarcely more charitable about the range of hills that lie behind the town, namely the Cotswolds. He found them bleak and desolate, and could think of only one part of England that was scenically less appealing – the blasted heathlands of Surrey.

However, today we think differently, and the Cotswolds are now officially designated an Area of Outstanding Natural Beauty. One of their great attractions is the combination of broad vistas over upland pasture with large winding valleys in which the famous stone-built villages lie. The Cotswolds are a simple escarpment, with the scarp face overlooking the patchwork quilt fields of the Severn lowlands, and the dip slope trending southeastwards

into the valley of the Thames. One of the most remarkable features of the dip slope is the great size of the valleys that form the headwaters of the Thames in comparison with the diminutive size of the present streams flowing in them. The streams flow in large, wide, flat-bottomed valleys meandering across the dip slope in great curves, called valley meanders. The Dickler, Windrush, Evenlode, Churn, Coln and Leach rivers and the Sherborne Brook all display this feature.

A great American scientist, William Morris Davis, was very struck by these large valleys and their small streams, which he called misfit streams. He believed that they were now so small because their source of water had become diminished by river capture (see sites 24 & 61); that is, some of the tributaries of the Thames had been captured by the more aggressive streams of the Severn system. However, more recent studies have shown that such valley meander systems and their misfits are very widespread, and that they occur both on the dip slope and

on the escarpment, a fact that cannot be explained by Davis's ideas about capture. Thus a more general process is required to explain misfits.

It is now believed that they result from the climatic changes of the Ice Age. Such changes have often been put forward to explain the dry valleys of the chalk lands (see site 44), and the valley meanders differ only in that they have greatly shrunken streams rather than being totally dry. By comparing the shape and size of the valley meanders and the present stream-channel meanders it is possible to arrive at some measure of the difference in stream discharge between that which carved the former and that which shaped the latter. The channels associated with the valley meanders are approximately ten times as wide as the existing stream beds, and the wavelengths of the valley meanders are approximately nine times as long as the wavelengths of the corresponding stream meanders.

How do we account for these

The flooded Leach Valley near the viewpoint, showing strip lynchets on the near slope

vastly greater river discharges in the past? One possibility is higher rainfall levels. Alternatively, when temperatures were lower during the glacial phases of the Ice Age, much precipitation would have fallen as snow rather than as rain. Rapid melting of the snow in the spring thaw would supply large discharges. In addition the amount of water running off the surface might have been increased by the presence of frozen subsoil (permafrost) which would have hampered the percolation of water into the permeable limestones.

One of the most pleasing places to see the valley meanders is on the River Leach near Sheepbridge Barn (SP 191068), for not only are there beautifully twisting valley curves, but there are also some ancient field systems, called strip lynchets, on the valley sides. They occur as a series of well defined terraces that were produced in times past by long continued ploughing on steep slopes.

Sometimes the River Leach is completely dry. The reason for this is that it now carries insufficient discharge for much of the year to be able to withstand percolation losses into the well jointed oolitic limestone across which it flows. In really dry years the Leach disappears underground at Larkethill Wood (SP 140116) only 7 km from its source. During most periods of low rainfall it goes underground for a distance of about 14 km, only reappearing in some springs and boggy ground just upstream from the charming village of Eastleach Turville (SP 201052). Thus, by standing at Sheepbridge one will often see a completely misfit stream; the stream channel dry, but lying in a major valley. By following the Leach down to Eastleach Turville one will come across a lovely example of a clear Cotswold stream flowing through a village that is best visited at daffodil time.

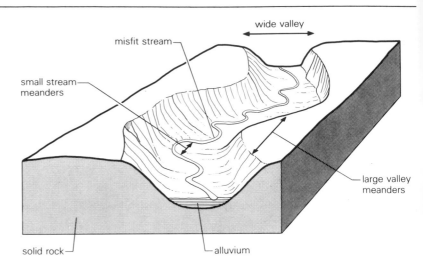

Characteristics of a misfit stream

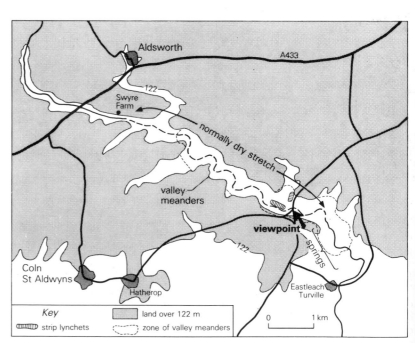

The location and form of the valley meanders of the River Leach across the dip slope of the Cotswolds

Viewpoint: grid reference: SP 193068 (Sheepbridge)
Map sheet 163

44 The Manger of the White Horse, Uffington

Looking from the top of the Manger towards the Vale of the White Horse, showing the chute-like features on the far slope

To the delight of Oxfordshire but the chagrin of Berkshire, the local government reorganisation of the 1970s transferred the famous White Horse of Uffington to Oxfordshire. The horse itself is a major attraction of great antiquity, although no-one is quite sure when this galloping creature was cut into the chalk escarpment of the Berkshire Downs. It looks down onto the vale to which it has given its name, and to its rear is the great prehistoric route – the Ridgeway – and an Iron Age fort. The attractiveness of the site, however, owes much to the setting, for at this point the escarpment is particularly high and steep, and peculiarly furrowed by a deep embayment, the Manger.

The Manger (SU 299867) is a fine example of one of those widespread phenomena of the chalk lands of southern England – the dry valley. The escarpment is breached by several such features between Wantage and Swindon, and they appear to be the result of more intense stream runoff in times past. Many ideas have been put forward to explain their development, most of them involving some fairly severe change of climate in the Ice Age. Some valleys probably formed when the water table was higher, so that spring activity was more intense. The springs may have carved their way backwards along joints in the chalk, dissolving, weakening and undermining the rock in the process (called 'spring sapping'). Alternatively, when conditions were especially cold, and the ground surface was underlain by frozen subsoil (permafrost), the spring-time snow melt and summer rain would not have been able to percolate downwards into the chalk, as it would today, but would run off the surface and erode into the chalk to produce a valley. Certainly, the chalk

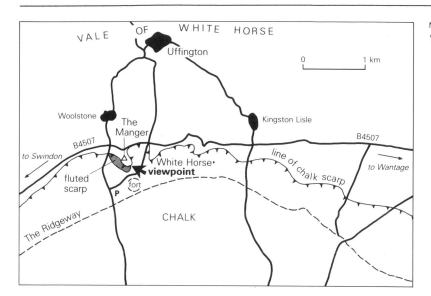

Map sheet 174

Viewpoint grid reference: SU 299867

appears to have been severely shattered by frost activity during the later phases of the Ice Age, and great spreads of chalk debris produced by this process and moved by sludging down hill (solifluction) occur at the foot of the scarp as a series of large fans.

The Manger, however, is unlike most other dry valleys, for on its southern flank it is serrated by a series of steep, narrow, chute-like features that give a washboard appearance to the escarpment. Some equally impressive features occur further to the west at Kingstone Combes (SU 273854). These chutes are some 120 m long, some 40 m across and about 45 m apart. They have steep slopes. There is no really satisfactory explanation for their development, but studies in present-day high-altitude and high-latitude areas, such as the Scandinavian mountains, suggest that similar forms are widespread, and that the chutes represent the tracks of avalanches. When avalanches occur they not only pose a threat to life, demolish houses and flatten trees, but they can also cause erosion of the underlying slopes. As a 'track' is formed by the avalanche in one winter the same route is more likely to be followed in a subsequent winter. In time, therefore, the avalanches produce a more and more marked series of furrows on a hill front. It is possible that the chutes we see on the side of the Manger developed in the Ice Age when masses of snow swept over the top of the chalk escarpment and carved out the remarkable series of parallel furrows.

45 Piggledene and Lockeridge

Early travellers, including Daniel Defoe and Samuel Pepys, passing along the main London–Bristol road in the vicinity of Marlborough were fascinated by one of the most curious phenomena of the English chalklands – the Sarsen stones or grey wethers. John Aubrey, for example, wrote:

'the stones called the Grey Wethers; which lye scattred all over the downes about Marleborough, and incumber the ground for at least seven miles diameter; and in many places they are, as it were, sown so thick that travellers in the twylight at a distance take them to be flocks of sheep (wethers) from whence they have their name. So that this tract of ground looks as if it has been the scene where the giants had fought with huge stones against the Gods, as is described by Hesiod in his Θεολοξια.

They are also (far from the rode) comonly called Sarsdens, or Sarsdon stones … They peep above the ground a yard or more high bigger and lesser. Those that lie in the weather are so hard that no toole can touch them … Many of them are mighty great ones, and particularly those in Overton Wood. Of these kind of stones are framed the two stupendous antiquities of Aubury and Stone-heng.'

The origin of the word 'sarsen' is still shrouded in mystery. A common theory is that it is derived from Saracen, while another is that it is derived from the Saxon *sar stan* (a troublesome stone) or *sel stan* (a great stone). No less taxing a problem than the origin of the word is the origin of the sarsens themselves.

Early ideas about origin include Sir Christopher Wren's suggestion that they were volcanic, Stukeley's

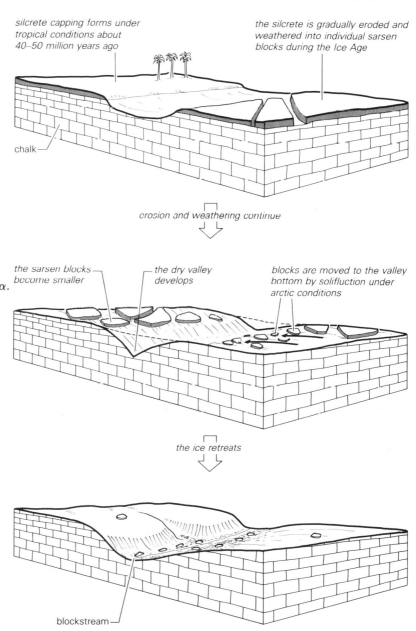

The formation of a sarsen blockstream by the disintegration and transport of a silcrete capping

silcrete capping forms under tropical conditions about 40–50 million years ago

the silcrete is gradually eroded and weathered into individual sarsen blocks during the Ice Age

chalk

erosion and weathering continue

the sarsen blocks become smaller

the dry valley develops

blocks are moved to the valley bottom by solifluction under arctic conditions

the ice retreats

blockstream

121

that they were expelled from the chalk by the rotation of the Earth, and another that they were scattered over the Downs by an explosion accompanying an earthquake. A local topographer, Adams, believed that marine action due 'to some convulsion of nature' was responsible for their scattered distribution:

'they are the waifs and strays of an appalling wreck, and their condition is akin to that of shipwrecked mariners on some foreign shore.'

The current explanation for their formation is as follows. About 40–50 million years ago, during the Tertiary geological era, gravelly and sandy beds were deposited on top of the chalk. At that time, or shortly after, the climate was tropical, perhaps like the Kalahari Desert of southern Africa today. Under such conditions some of the quartz sand may be dissolved and the silica deposited nearby in the sands as a cement, binding the remainder of the sand and gravel together. This produces a new hard surface or capping comprising sheets or discontinuous masses of silica (termed 'silcrete').

Subsequently, during the rigorous tundra conditions of the Ice Age, the silcrete was broken up by frost action into individual blocks (the sarsens). Moreover under these harsh conditions the underlying chalk was readily weathered, producing a thick mantle of rock debris on the slopes. This debris, especially when saturated by spring-time snow melt, crept or sludged down to collect in the valley bottoms, by the process called solifluction. The same sludging occurs in Arctic areas today and seems capable of moving large boulders. It was probably this process that transported the sarsen blocks down into the Kennet and its tributary valleys to give the so-called 'blockstreams'. Similar features are seen near Portesham in Dorset, and further information about possible modes of origin is given in the description of that site (52).

These blockstreams, composed in places of hundreds of blocks weighing as much as 40 to 200 tons each, are widespread in the Marlborough area. They were even more extensive before many of the

The sarsen blockstream at Lockeridge

blocks were removed to build prehistoric monuments such as neighbouring Avebury, or cut up to make building blocks for Windsor Castle. Two accessible National Trust areas where they can be admired in a beautiful setting are Piggledene (ST 142686; to the north of the A4) and Lockeridge (ST 146674; to the south of it). Lockeridge village, which must be passed through to reach its blockstream, is notable for the many examples of domestic architecture in which the sarsens have been used.

The traditional use of sarsens in local buildings

Map sheet 173
Grid references: Piggledene ST 142626
Lockeridge ST 146674

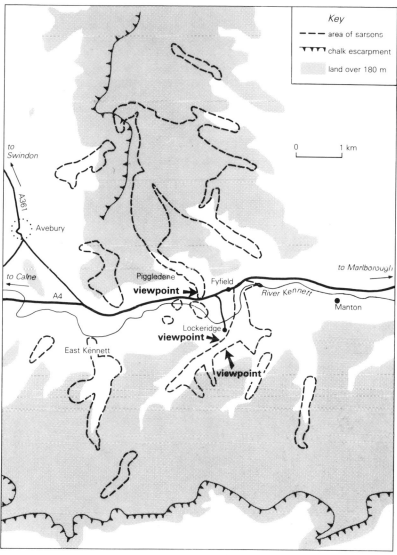

46   Clifton Gorge and the anomalous course of the Avon

There are very few cities that can boast a classic landform right in their heart. Edinburgh's famous castle rests on an ancient volcano, Durham's cathedral lies on an enormous meander core of the river Wear, while Bristol has in its suburb of Clifton the spectacular gorge spanned by Brunel's elegant suspension bridge.

The Clifton Gorge (ST 564730), excavated by the River Avon, is cut into a plateau of Carboniferous Limestone, the rock that forms the Mendip Hills nearby (site 47). The Avon starts as a small stream in the Cotswold hills near Badminton and from there on follows a bizarre course. First it turns eastwards to Malmesbury, and then it turns south and runs through Chippenham to Bradford-on-Avon. Before reaching Bath it takes a right-angled bend to flow through the limestone escarpment in a gorge at Limpley Stoke and later at Clifton. Thus, the Avon passes from one lowland to

another through deep gorges cut in the limestone escarpments. There would appear to be many other courses along which the Avon might have flowed more easily, so why did it choose this peculiar route?

The earliest explanation for the gorge comes from a local legend that two giants, Vincent and Goram, wanted to leave a permanent memorial to themselves. They decided to do this by each digging a ravine – Vincent at Clifton and Goram at Henbury – so diverting the course of the Avon.

When the science of geology developed in the 19th century, such explanations were no longer regarded with favour, and more prosaic interpretations were adopted. That aristocratic geologist, Sir Roderick Impey Murchison, believed that the gorge was the result of a massive, convulsive dislocation of the area, producing in effect a great rent in the landscape. This idea was later dealt with scathingly by another

distinguished Victorian geologist, A. C. Ramsay:

'The vulgar notion respecting the Avon and its gorge is, that before the ravine was formed all the low ground through which the river and its tributaries flow was a large lake, that "a convulsion of nature" suddenly rent the rocks asunder and formed the gorge through which the river afterwards flowed, and so drained the hypothetical lake. It is scarcely necessary to add, that had there been a large lake in the area, we might expect to find lacustrine deposits and organisms in some parts of these valleys, but none exist.'

Nonetheless the presence of a lake to explain the gorge is postulated even now by those who believe that, like so many other landforms in England and Wales, it is a product of the Ice Age. The argument runs as follows. A former course of the Avon was blocked by ice at some stage during the Ice Age, and the river ponded up behind the ice to form a lake. Glacial melt water added to the volume of the lake, which eventually grew so large that it overflowed, cutting a new route to the sea and carving the Clifton Gorge and other 'canyon-like valleys' in the process (see also site 57). Once this new route was established, the Avon continued to flow along it even after the disappearance of the ice sheet because it was lower than the old route taken by the river.

For many years, however, there was little evidence that the area had been extensively glaciated, or that a

St Vincent's Rocks, Clifton.

A 19th-century engraving of the Clifton Gorge

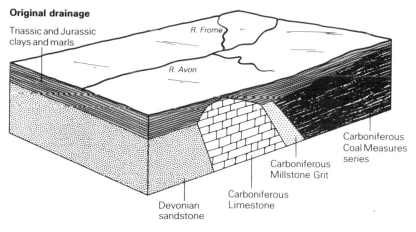

Original drainage

Triassic and Jurassic clays and marls

R. Frome

R. Avon

Carboniferous Coal Measures series

Carboniferous Millstone Grit

Carboniferous Limestone

Devonian sandstone

younger rocks are stripped by erosion

The same drainage pattern superimposed on older underlying rocks

thin remnants of Jurassic and Triassic clays

Clifton Gorge is cut through the limestone

R. Frome

R. Avon

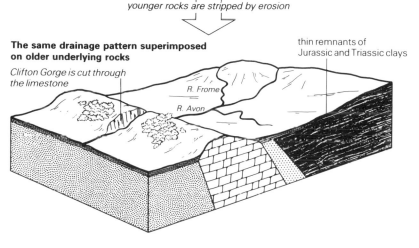

The formation of the gorge at Clifton as a result of superimposed drainage

lake had existed (shown by lake sediments or old shorelines). Nevertheless, recent work has suggested that ice sheets moving across from Ireland and Wales passed over the area during the Ice Age. If this is so, the various peculiar drainage lines, of which the Clifton Gorge is but one, could have resulted from erosion by glacial melt water.

The alternative explanation, and probably the most acceptable, is that the gorge is an example of 'superimposed drainage' (as in sites 41 & 16). It is envisaged that the Carboniferous Limestone and older rocks of the area were once covered by a layer of younger, Jurassic and Cretaceous rocks. A drainage system established itself on these younger rocks. Over the years, erosion by the Yeo and Frome rivers and Colliter Brook led to the gradual stripping away of the younger rocks, revealing the older Carboniferous Limestone beneath. Since this was a gradual process, the course of the river did not change, and so it became superimposed on the older rocks. Thus, the drainage pattern had formed in response to the structure of the younger rocks and was superimposed on the very different structure of the older rocks. Gorges were formed in the more resistant of the older rocks, as the river cut down through them.

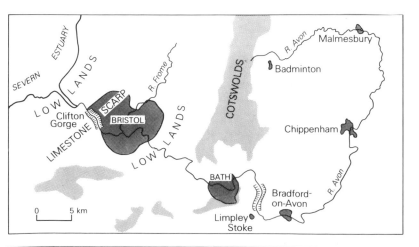

The peculiar course of the River Avon

47   The Cheddar Gorge: 'a frightful chasm'

'A deep frightful chasm in the mountain' is how Daniel Defoe described the Cheddar Gorge in 1724. Since then millions of visitors have come to gaze in awe and wonder at the most renowned landform in south-west England.

The 'mountain' into which the gorge slices is the Mendip Plateau, a relatively high area composed predominantly of Carboniferous Limestone with a core of Old Red Sandstone. Chemical solution of the limestone gives rise to a whole series of characteristic features (termed 'karstic' after the Kars limestone area of Yugoslavia). Acidic waters running off the sandstone core dissolve the calcium carbonate which makes up the limestone. Solution and the presence of many joints in the limestone encourages much of the drainage to go underground, widening the joints in the process.

In the 18th and 19th centuries the Mendip gorges were thought to have resulted from catastrophic happenings associated with earthquakes and the like. Subsequently, a popular explanation was that of an unroofed cavern. It was believed that a very large underground chamber developed as a result of the solution of the limestone on a massive scale. When this chamber reached a critical size, the weakened roof collapsed, forming the gorge. However, this idea is now largely discredited, for the cavern would have had to be taller and wider by far than any known in Britain, and there is no evidence of a substantial quantity of debris from the collapsed roof.

It is now believed that the Cheddar Gorge and its associated dry valleys were cut by surface streams. The upper, gentle part of the gorge was cut about 2–3 million years ago (late Pliocene and early Pleistocene) before drainage had gone

underground. The lower, steep-sided parts were caused by intense stream erosion during glacial phases of the Ice Age, when temperatures were so low that the ground surface was permanently frozen. The freezing of water within the limestone rendered it impermeable, and this forced the drainage to run across the surface of the rock, rather than underground.

There are various pieces of evidence supporting the idea that

river erosion produced the gorge. First, the spurs that make up the side of the gorge interlock, just as in a normal river valley. Secondly, the collecting area of the valley system leading into the gorge is many times greater than those leading into the other, smaller gorges in the Mendips. The greater amount of water funelled through Cheddar helps to explain its size. Thirdly, some river erosion still takes place after heavy

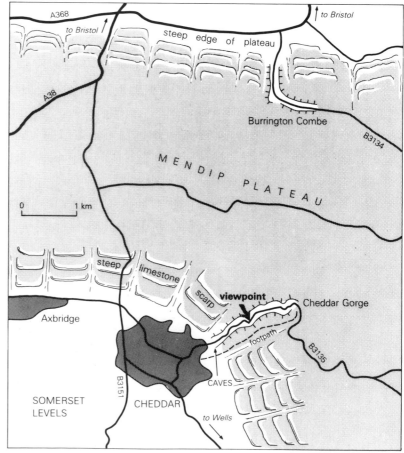

Map sheet 182
Grid reference: ST 475544
Roadside parking areas throughout the gorge
Shop, show caves, etc.

rainfall. In the great flood of July 1968, for example, 125 mm of rain fell in a few hours, and a large river cascaded down the gorge, flooding and damaging property in the area.

While driving through the gorge you may see that it is not symmetrical. This is because the shape of Cheddar cliffs is largely controlled by rock structure. The limestone layers generally dip southwards, while the gorge itself runs almost east to west. Thus the northern cliff face is gentler and it slopes at an angle similar to that of the tilted layers in the limestone. On the other hand, the southern cliff face is often near vertical.

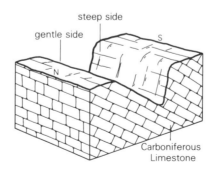

Cheddar Gorge from the west, showing the geological control of the valley asymmetry

The gorge can be viewed from the B3135, which passes through it. However, the best view of its precipitous western section and of the Somerset Levels beyond can be had after a brisk climb up the 250 or so steps of Jacob's Ladder, opposite the Cliff Hotel. There is a great deal to be said for distancing oneself from some of the tourist trappings, and the view from above transforms one's perspective on the gorge. Instead of thinking how narrow it is, one is impressed by how enormously wide it is at the top – one more reason for doubting that it was ever a gigantic cavern.

48 The Needles

The thin ridge of chalk in the Isle of Wight and its association with The Needles, Old Harry and the Isle of Purbeck

The Needles have always been one of the great scenic attractions of the Isle of Wight, a fact recognised by an early 19th century topographer, Sir Henry Englefield:

'Nothing can be more interesting, particularly to those who take pleasure in aquatic excursions, than to sail between and round the Needles. The wonderfully coloured cliffs of Alum Bay, the lofty and towering chalk precipices of Scratchell's Bay, of the most dazzling whiteness and the most elegant forms, the magnitude and singularity of the spiny, insulated masses, which seem at every instant to be shifting their situations, and give a mazy perplexity to the place, the screaming noise of the aquatic birds, the agitation of the sea, and the rapidity of the tide, occasioning frequently a slight degree of danger, all these circumstances continue to raise in the mind unusual emotions, and to give the scene a character highly singular, and even romantic:'

In times past they may have been even more singular and romantic, for their name is derived chiefly from one of their number, much taller and more slender than any of those now remaining, which fell suddenly under the onslaught of the sea in 1764. The Needles and the neighbouring cliffs were at that time frequented by vast swarms of sea birds – puffins, razorbills, gulls, cormorants, choughs, and many others. The local people were in the habit of catching them by the hazardous method, also practised in the Shetlands and the Faroes, of being swung over the brow of the cliff by a rope made fast in the earth above. It is recorded that in 1781 the soft feathers obtained from a dozen birds were sold for 8 pence; and the carcasses were then disposed of, at the rate of a halfpenny each, to fishermen, who used them to bait their crab-pots.

The Needles are composed of the same white chalk, seamed with rows of dark flints, that makes the Downs of south-east England. In the Isle of

Wight, the chalk forms the narrow ridge between the Needles in the west and Culver Cliff in the east. Close inspection will reveal that the layers of chalk have been folded almost vertically, forming part of an upward fold (or anticline). This is well seen in the cliffs to the south of the Needles. The folding took place at the same time as the Alps were formed in Europe.

The relative resistance of the chalk, compared to other rocks in the area, is shown by the height of the ridge and in the fact that the longest dimension of the island is along the outcrop of the chalk, east to west. Thus, the chalk forms prominent headlands at the coast, whereas the softer clays and sandstones to the north (Tertiary age) and to the south (Jurassic age) have been eroded into broad bays.

Where the chalk ridge has come under marine attack on the exposed western headland, it has been fragmented, the remnants forming a set of jagged stacks called the Needles. Eventually, the Needles will succumb to the onslaught of the waves, leaving a few stumps of eroded rock in their place. Indeed, depending on the state of the tide you may see a few exposed stumps of chalk that just manage to poke out above water level. These are testimony to the former existence of more numerous Needles, such as the one that collapsed in 1764. The best way to view this detail (and the lighthouse perched on the end of the outermost stack) is from a boat, and trips are often available in the summer. From the boat you can also pick out the high water mark – the top of the discoloured area around

the base of the stacks – rather like a scum ring around the bathtub.

The view from the cliff above Alum Bay will confirm that the chalk ridge of the Isle of Wight is in line with that on the Isle of Purbeck, just north of Swanage. At one time the Isle of Wight was part of the mainland, and has since been separated by erosion. One of the best views of the Needles from the land is also obtained from the same point. There is ample car parking nearby and even a chairlift to take you down to the beach and the coloured cliffs. These Tertiary beds in Alum Bay also exhibit nearly vertical folding, like the chalk, but their rich and varied colours provide a dramatic contrast with the white chalk. Unfortunately, the unique cliffs are currently being damaged by the collection of their coloured sands by

The Needles at the western end of the Isle of Wight are among Britain's most famous stacks

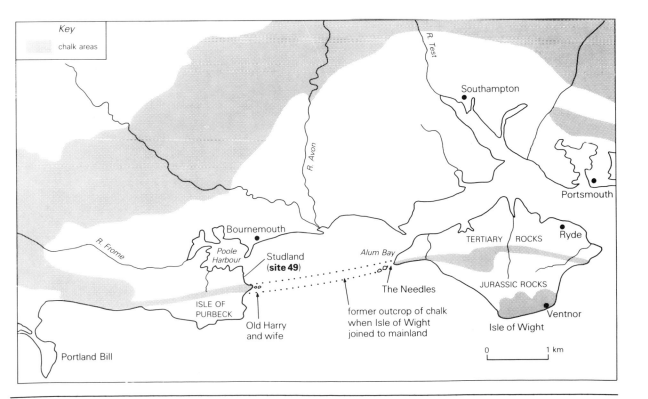

129

tourists and by the hammering of parties of geologists.

While on the Isle of Wight it is also worth looking at the massive landslips around Ventnor and Blackgang in the south-east. Only recently a slide at Blackgang destroyed a road and holiday cottages. The slumped mass forms a distinct second cliff ('undercliff'), which then protects the original cliff from further wave attack. Sliding and slumping is encouraged when the cliffline is undermined by wave erosion; when the rocks dip towards the sea; and when there is a build-up of water above layers of clay (Gault Clay), as described for parts of the Dorset coastline (site 55).

Map sheet 196
Viewpoint grid reference: SZ 305854
Chairlift, shops and café

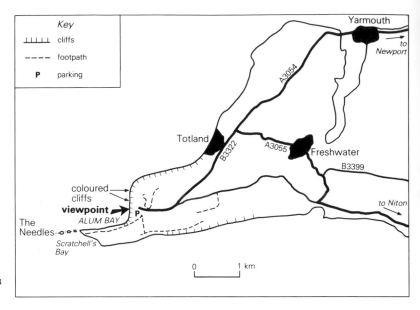

49 Studland Heath and the Little Sea

South of Poole Harbour, lying between it and the sea, is an area of dunes, lakes and marsh called Studland Heath. The dunes are some of the best on the south coast, forming four main ridges, one of which (No. 4) has developed since a comprehensive survey made in the 1930s. The earlier ridges formed between 1607 and 1721 (No. 1), 1721 and 1849 (No. 2) and 1849 and 1933 (No. 3). They all trend a little east of north, parallel to the coastline, and are separated by low marshy tracts and ponds. These are typical foredunes, formed on the beach, at the shoreline, and blown a short distance inland. The white quartz sand making up the dunes was probably combed in from the sea floor by the rising seas in the early part of the Holocene. (Much sand was deposited on the sea floor by rivers at times of low sea level in the glacial phases.) Later on the waves washed the sand onto the beach.

The older ridges, which have been leached by rain water for longer, have developed a more acidic soil than the freshly deposited sand areas, though the Studland dunes are curious in that they contain relatively little lime. The soils on the youngest ridge are only slightly alkaline (with a pH of 7.5). The soils in turn affect the vegetation, and so the youngest ridge, immediately behind the wide, excellent beach, is mostly covered by sea-water tolerant and alkaline-loving marram grass. This helps to stabilise the dune but it gradually dies out as the soil changes and other vegetation colonises. The inner sand ridges are dominated by heather (*Calluna*), bell heather (*Erica*), gorse (*Ulex*) and some bracken. The acidity is generally high in the poorly-drained troughs or 'slacks' between the dunes. High acidity in the slack

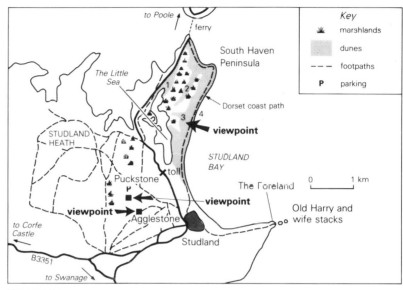

Map sheet 195

Viewpoint grid reference: dunes SZ 036846   Agglestone SZ 023828   Puckstone SZ 021831

behind the youngest ridge has made this a favoured site for the reed *Juncus maritimus*.

A large shallow lake, the Little Sea, is also found within the heathland. This has a well documented history. In 1721 it was a tidal inlet open to the sea; by 1785 it was changing into a brackish lagoon; and by 1900, as it was sealed off by dune chain No. 3, it became a freshwater lake, and so it remains today. The Little Sea covers about 28 ha but has an average depth of only 1 m, due to infilling by sand and muds. It has a floor of clean sand with local patches of mud, and the bottom is relatively level.

Other curiosities in the heathland area are two masses of cemented Tertiary beds of Eocene age: the Agglestone (SZ 024828) and the Puckstone (SZ 022831). They are formed of the iron-cemented Agglestone Grit, remnants of which litter the Studland Heath area.

Measuring approximately 8.5 m by 6 m by 4.5 m, the Agglestone is the largest remnant of this bed. It resembles an inverted cone or anvil resting on a 21 m high mound of softer, Tertiary, Bagshot Sand. The Agglestone has partially protected this mound from erosion and weathering. In the 18th century the Agglestone was called the 'Devil's Nightcap' and tradition has it that the Devil threw it from the Isle of Wight with the intent of demolishing Corfe Castle, but that it fell short. Only about 350 m north-west of Agglestone is the much smaller Puckstone, a lichen-covered rock about 1 m high on its protected mound about 15 m in diameter.

While in the area it is worth visiting the spectacular chalk cliff scenery to the east of Studland. Originally the chalk extended across the sea to join up with that on the Isle of Wight (site 48). Since then it has slowly been worn away by solution

The Little Sea enclosed by the vegetated dune ridges of the Studland Peninsula

and the battering of the waves. Today, two rather splendid stacks, isolated from the mainland cliff by marine action, can be seen off The Foreland (SZ 055825). The Dorset Coast Path follows the top of the cliff – an invigorating walk.

50 Lulworth Cove

A view of Lulworth Cove from the ridge behind Stair Hole

Lulworth Cove and its neighbouring bays are so popular that the car parks often cannot cope with the volume of visitors, and footpath erosion is becoming an increasing problem. Some visitors come here to enjoy the scenery, others to examine the rocks and fossils, and some to try and fathom out how this remarkable coastline evolved. These sentiments are well expressed by Woodward, who said, in 1908,

'Its surprising renown is probably due to the fact that its attractions are threefold in character; one might almost say they symbolise the good, the true and the beautiful. They certainly represent utility, science, and beauty; a safe harbour, geological evidences and exquisite scenery.'

Traditionally, the cove is said to illustrate an important principal: that stage of development is an important consideration in understanding landforms. It is impossible to observe landscape evolving from its beginnings to a fully developed form because the shaping processes of weathering, erosion and deposition are too slow. However, sometimes a sequence of features occurs in space which, if they are formed under similar conditions, can be arranged in a time-based sequence from oldest to youngest. The coast between Durdle Door and Worbarrow Tout may be such a sequence.

It is believed that Stair Hole represents the beginning of a new bay, that Lulworth Cove is an intermediate but beautifully developed feature that is as near

perfect as can be conceived, and that Hope Bay, Worbarrow Bay and Man o'War Cove at Durdle Door represent the final condition where the original geological controls have all but been destroyed and where a new system comes into play.

The argument to explain this sequence is that the sea fights to penetrate the coastal barrier of Portland and Purbeck Limestone, and breaches it in narrow arches which collapse to form entrances to small bays such as at Stair Hole. Thomas Hardy describes the projecting rocks on either side of the aperture as 'The pillars of Hercules to this miniature Mediterranean'. Then when the sea reaches the soft Wealden Beds of sands and clays behind it, it causes rapid erosion so that a wide, near-circular bay is scooped out behind the narrow entrance in the limestones, as at Lulworth Cove. The beauty of the cove is enhanced by the backdrop of steep white cliffs of chalk, which the sea has difficulty in eroding. Thus, the development of the bay becomes slower than the widening of the mouth, because of the difference in ease of erosion of the various rocks. So, the Portland and Purbeck Beds are destroyed progressively until only small stacks remain, such as the jagged teeth of Man o'War Rocks and Mupe Rocks. At this point shore erosion becomes very important and the bay widens right out at its mouth (e.g. see Worbarrow Bay) perhaps to merge with neighbouring bays and to form a straighter coastline of chalk cliffs. St Oswald's Bay is an elongated double bay formed by the amalgamation of two former coves comparable in form and origin (if not scale) with Lulworth, and the coast between Durdle Door and Bat's Head appears to be the result of the amalgamation of up to three coves. These represent the final stages of the sequence in which the Purbeck Limestone barrier has almost disappeared, with only the Bull, Calf and Cow rocks exposed at low tide.

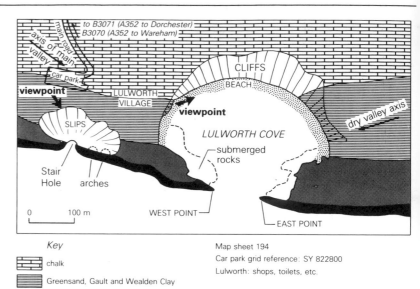

Key
- chalk
- Greensand, Gault and Wealden Clay
- Purbeck Beds
- Portland Beds

Map sheet 194
Car park grid reference: SY 822800
Lulworth: shops, toilets, etc.

However, marine erosion is not the only process involved in cove formation. Several other factors must be taken into account. All of the bays are fed by large valley systems, some of which carry streams that flow all year. These valleys are much too large to have been formed after the breaching of the coastal limestone barrier and therefore they must have flowed out to sea through a valley – a breach – in the Portland and Purbeck

The Dorset coast between Durdle Door and Worbarrow Bay – geology and relief

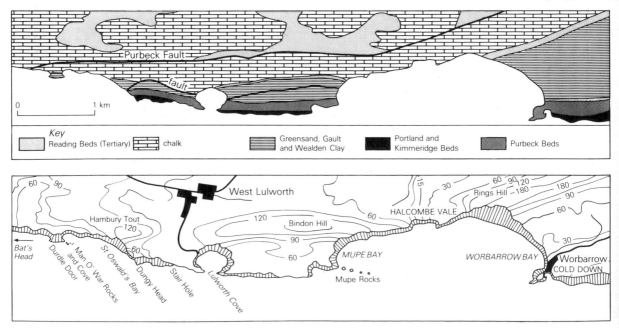

Key
- Reading Beds (Tertiary)
- chalk
- Greensand, Gault and Wealden Clay
- Portland and Kimmeridge Beds
- Purbeck Beds

stone. In other words, it is not necessary for *marine* breaching to occur before a bay can develop. Stair Hole is unique in that it does not have a valley system and so perhaps it should be considered as a completely separate case. Taking this view further, Lulworth Cove (and also probably Worbarrow Bay) merely represent partially drowned river valleys.

A further point is that the streams would have flowed to the lower sea levels of the glacial periods. For example, the water in the entrance to Lulworth is about 16 m deep. This low base level combined with the greater runoff under arctic conditions (see site 47) would have increased the erosional power of the streams, making the opening of a deep gorge through what is now the coastal barrier a simple matter. At this time the Chalk and Wealden Beds formed valley side slopes covered in weathered debris and affected by solifluction and other types of mass movement. As the sea level rose in postglacial times, it flooded these valleys and trimmed the lower part of the cliffs within the cove into the perfect shape we know today. This was assisted by rapid mass movement in the Wealden Beds, the present slopes of which are scored by many slumps, gullies and mudslides.

Lulworth is reached by the B3071 from Wool or the B3070 from Holme Bridge (SY 891870). Use the car park at SY 822800; direct vehicular access is not permitted to the cove itself. There are signposts to Stair Hole and Lulworth and a footpath to Durdle Door over Dungy Head high above St Oswald's Bay. The cliffs in the area are precipitous and unstable. Keep well away!

Man O' War Rocks and Cove from Durdle Door

51　Culpepper's Dish

Thomas Hardy, the Wessex novelist, immortalised Puddletown Heath as the famous 'Egdon Heath'. The heathlands of Dorset and Hampshire that so fascinated Hardy occupy a lowland basin bounded on the north by the rolling chalk downlands of Cranbourne Chase and on the south by open grassland plateaux and ridges of the Isle of Purbeck. To the layman, the sands, clays and gravels underlying the heaths scarcely merit the term 'rocks', but they were nonetheless laid down as some of Britain's youngest rocks, within the past 70 million years. Their acid soils are infertile and are thus much given over to military use, to forestry, and to the growth of heather and gorse.

One of the most intriguing landscapes in the heathland is the group of hollows clustered near Puddletown. The unusual nature of the heathland around Puddletown was recognised by Stevenson in his *General view of the agriculture of the county of Dorset* (1812):

'The heaths near Piddletown, and six or seven miles further to the south-east, are remarkable for the many round deep pits which they contain; they diminish almost to a point at the bottom, and are not unlike inverted cones. There is one which appears to be 20 yards wide, and 10 yards deep. They are said to become deeper, and an idea is entertained, that they are undermined by concealed streams of water.'

The Dorset hollows are concentrated on heathlands between Dorchester and Bere Regis, between the Rivers Frome and Piddle, covering an area of 16 square kilometres. The main group of more than 370 hollows is on Puddletown Heath (SY 740925) where Tertiary Reading Beds and Plateau Gravels overlie the Upper Chalk. A second group of more than 100 hollows is on Southover (SY 785925) and Affpuddle (SY 804923) Heaths. Still more hollows are found much further west on Black Down (SY 612877), near the Hardy Memorial, where Bagshot Beds 10 m thick cover the Upper Chalk over an area of 2.5 square kilometres (see site 52).

Most of the hollows are 10–20 m wide, but Culpepper's Dish, probably named after the London apothecary, is atypically large with an average diameter of no less than 86 m. Some other very large hollows are found nearby. The hollows are approximately circular, though compound depressions occur when a

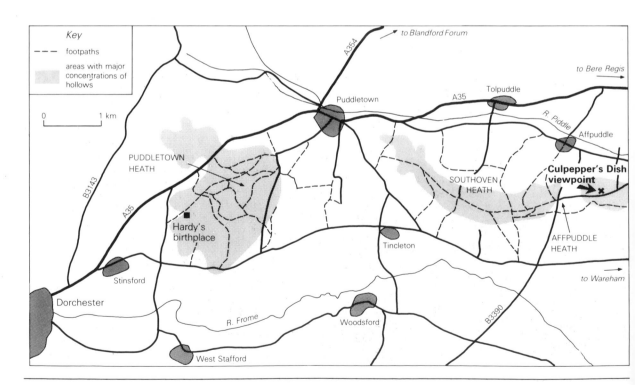

By solutional widening of joints in the chalk

By collapse of chalk and Tertiary beds into a chalk cave

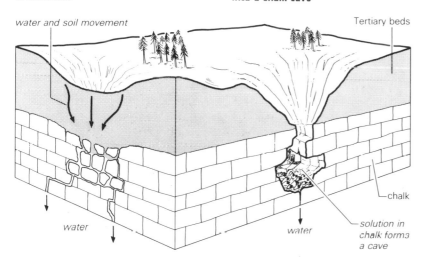

Two possible mechanisms by which the hollows of the Dorset heathlands may have formed

Culpepper's Dish – a 21.2 m deep hollow formed in the chalk

small hollow has collapsed within a larger one (e.g. at SY 737931), and sometimes an elongate form may result (e.g. at SY 791923). Most of them are relatively shallow, averaging 2–4 m, but once again Culpepper's Dish is atypical, having a depth of 21.2 m. The hollows have moderately steep slopes. One particularly striking characteristic of these various forms is their great density; this reaches 99 per square kilometre on Southover Heath and 157 per square kilometre on Puddletown Heath.

How can we account for the profusion and location of all these depressions? The explanation probably lies in the fact that the chalk is overlain by a variable, but sometimes great, depth of superficial material, predominantly of Tertiary age. This is composed to a great extent of sands, flints and gravels, with lesser amounts of clay, and it supports acidic vegetation rather than the alkaline soils and vegetation characteristic of chalk. Water moving through the soil beneath this vegetation also becomes highly acidic and can thus dissolve the chalk.

Some of the hollows may originate simply by solution at the boundary between the chalk and overlying superficial material, but elsewhere the solution proceeds to a deeper level, leading to dramatic collapses of the ground surface into underlying cavities or fissures in the chalk. These are still forming at the present time. In the Bagshot Beds on top of Bronkham Hill (SY 624873), a hole about 0.3 m in diameter appeared on the surface, and beneath it was a large cave at least 6 m in diameter and about 6 m deep. This collapse probably occurred in 1956. A hollow at Black Down, now fenced off on account of the danger it presents to the unwary, appeared in about 1960 at a point 330 m away (direction of 320°) from the Hardy Memorial. When measured in August 1971 it had a remarkable beehive shape; 6.9 m deep, an opening only 2.7 m in diameter, and a maximum width inside of 3.9 m.

52 Portesham's rock-infested valley

The Valley of Stones, looking northwards from near the viewpoint

The Hardy Memorial (SY 613876) near Dorchester, which commemorates not the famous Wessex novelist, but the great sailor of the Nelson era, is at a good vantage point for some of the finest landscape in southern England. It is positioned where a capping of Tertiary gravels rest on the chalk. Exposures in these superficial gravels can be examined in small quarries around the car park at the monument itself. In some places they have led to accelerated solution of the underlying chalk and 'solution hollows' thus formed are numerous in the vicinity (see site 51). However, locally the flints and cobbles of the Tertiary beds have been cemented together to give sarsen stones of the type encountered in Wiltshire (site 45), elsewhere in southern England,

and in both the Netherlands and the Paris Basin, where they are called *meulières* (millstones).

Sarsens (also called 'grey wethers'), are deposits cemented by silica (quartz) to give durable lenses and blocks of tough sandstone. They are often composed of 98–99 per cent silica, and this tends to make them resistant to chemical weathering. In addition, they are not very porous and so they have tended to withstand the effects of frost action. Their appearance, however, varies greatly depending on the nature of the original material. They range from conglomerates with rounded pebbles, to breccias with angular pebbles and a high flint content, to nearly pure quartz sandstones. In the Portesham area the breccia and quartz types are dominant.

The origin of these materials is still obscure. The silica cement may have been derived from silica-rich groundwater, or in alkaline lake environments, or derived from solution of silica-rich dust blown from a desert. Similar deposits are a feature of many semi-arid tropical and subtropical regions today, especially the Sudan zone of Africa, the Kalahari and Namib deserts of southern Africa, and the arid interior of central Australia. They appear to be a type of hardened crust formed at or near the surface of the land ('duricrust'). So, it seems that dry tropical conditions may have existed over southern England at times in the Tertiary. The presence of tropical plant remains in Tertiary beds at Bovey Tracey in Devon lends support to this idea.

The sarsens are well displayed in the Valley of Stones (SY 596873), about 2 km west of the Hardy Memorial. Lines of large sarsen stones litter the area between Black Down Barn (SY 594873) and Little Bredy Farm (SY 594883), especially on the southern side of the valley, adjacent to the road. They are particularly well exposed where small seepage lines have removed the fine material which previously covered, or formed the matrix of, the sarsens. Within this area there are two main groups of stones: an upper and larger group oriented SW–NE below Black Down Barn (A) and a more scattered group further down valley (B). Some of the blocks (group B) appear to have moved considerable distances down valley from their most likely source – perhaps as much as 1800 m of movement over very gentle slopes. These blocks are aligned with their longest dimension down slope.

The blockfield of sarsen stones probably existed originally as a near-continuous sheet or lens. During the Ice Age, frost weathering and mass movement (solifluction) disrupted this sheet, producing the individual blocks which periodically sludged downslope (see site 45). These processes have altered the configuration of the valley quite considerably.

Man has also played an important role in affecting the detailed shape of the valley. Many ancient field systems have, in part, accentuated the natural breaks of slope. A general view of the sarsen stones, together with the ancient field systems and solution hollows near Black Down and Crow Hill, can be obtained from the Abbotsbury–Martinstown road at SY 596872.

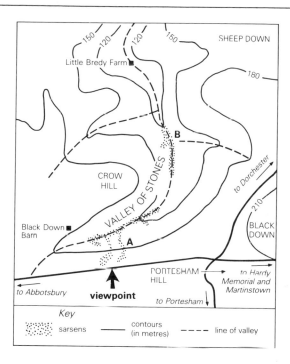

Map sheet 194
Viewpoint grid reference: SY 597871

A sarsen blockstream and ancient field systems in the Valley of Stones

53 Portland Bill: an abandoned beach

Although the hand of man rests rather heavily on large parts of 'The English Gibraltar' – the isle of Portland – it still undoubtedly has its charms. At its southern tip is Portland Bill (SY 675686), one of the most important Ice Age exposures along the whole coastline of southern England. Between the ugly military installation (to the west of the car park) and the sea cliff there is a relict of the time when sea level was higher than today – a raised beach.

During the Ice Age, as we show many times in this book, the climate fluctuated between warmer (called interglacials) and colder periods (called glacials). These changes were worldwide and had considerable impact on the levels of the oceans. When it was warm, perhaps even a few degrees warmer than today, the great polar ice caps shrank and so more water was available to fill the oceans. Conversely, when it was cold a great deal of water was stored up in ice caps which were three times more extensive than they are now. Consequently sea levels fell by more than 100 m.

The changing nature of climate and sea level can be reconstructed from the sequence of deposits at Portland Bill. At the top of the section is a deposit of 'head'. This is a highly angular mass of frost-shattered debris, most of which is from the local Portland and lower Purbeck Limestones. Along other parts of the island remains of mammoth, woolly rhinoceros and other mammals in this bed indicate, together with the nature of the deposit, that it was probably formed in sub-arctic conditions. The bed is generally about 1.0–1.5 m thick.

Beneath the head is a deposit of loam of approximately similar thickness, containing many shells of land snails (molluscs), the commonest being *Pupilla muscorum*. The loam was

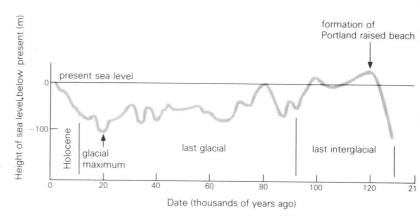

The curve of world-wide sea-level change illustrating the time of formation of the Portland raised beach: sea levels are high during warm interglacials, but low during cold glacials

probably deposited in a mere enclosed behind a storm beach after a slight retreat of the sea. The assemblage of shells is that of a temperate, or interglacial, climate similar to today.

Underneath these two deposits is a striking raised beach, at about 12 m above current sea level. The beach rests upon a platform cut in bedrock of Jurassic Portland Limestone, and the platform slopes from around 15 m to about 6 m above present mean sea level. Such a gradient is common on any recent wave-cut platforms developed where tidal range and wave activity were moderate.

The beach gravel above the

The Portland raised beach, looking from the viewpoint near Portland Bill towards the Ministry of Defence Establishment

140

platform consists dominantly of wave-rounded flint, local limestone and chert. Only about 1 per cent of the total is made up of 'foreign' material, such as quartzite and sandstones from the Budleigh Salterton Pebble Bed, tourmaline-rich rocks from Cornwall, and some pink granite. The beach also contains large quantities of shells, many well preserved, which indicate sea-water temperatures at the time of deposition slightly lower than those of today. In all the raised beach deposit is about 3–4 m thick, and it illustrates cross bedding and other interesting structures commonly observed in beach sediments. These structures mostly result from local variations in wave patterns and tidal currents. At the base of the raised beach there are many very large limestone boulders, often loosely bound together by natural cement.

Another interesting phenomenon, produced by changes of climate in the Quaternary, is found in the rock beneath the raised beach. The Purbeck Limestones, and sometimes the Portland as well, are much broken up and mixed – the result of pressures created by freezing of the ground surface. The depth to which they are affected may be as much as 3 m. At a number of locations, rubble from the lower beds has been churned up through the upper beds. It seems likely that all these features were associated with permafrost in a cold phase that preceded the formation of the raised beach. They are best displayed for a distance of 1.2 km NNE from the new lighthouse.

Thus the sequence as a whole represents colder glacial conditions, leading to rock shattering, followed by a warmer phase with higher (interglacial) sea levels, during which time the platform was cut, the beach was formed and the loam deposited. Lastly, the head at the top of the section indicates a reversal to colder conditions prior to the present temperate phase.

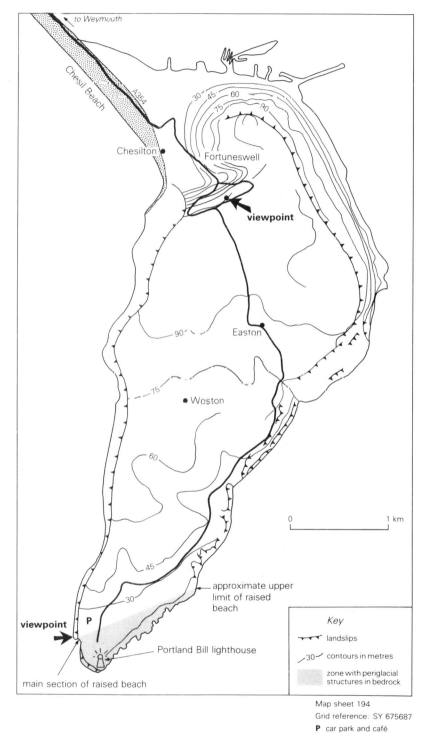

Map sheet 194
Grid reference: SY 675687
P car park and café

141

54 Chesil Beach and the Fleet

Chesil Beach is the most impressive shingle complex in the British Isles. Indeed, Lord Avebury described it as 'probably the most extensive and extraordinary accumulation of shingle in the world'. It is remarkable for its size, its regular crest line, its alignment, its lack of lateral ridges, and for the sorted material of which it is composed. In addition to this it is the only major shingle structure on the coast of Britain which is so long and straight. It is hardly surprising, therefore, that it has been the subject of some 60 or 70 published articles, and was described by many of the early topographers.

The Bank was, for example, noted by Leland:

'A little above Abbates-Byril is the hed or point of the Chisil lying north weste, that from thens stretch up 7 miles as a maine narow banke by a right line on to the south est, and then buttith on Portland scant a quarter of a mile above the new castell in Portland.'

It was also commented upon by Defoe, in 1724:

'Tho' Portland ſtands a league off from the main land of Britain, yet it is almost joyn'd by a prodigious Riſſe of Beach, that is to ſay, of small stone caſt up by the Sea ... When ships coming off from the Weſtward omit to keep a good offing, or are taken ſhort by contrary winds ... if they come to an Anchor, and ride it out, well and good, and if not, they run on shore of that vaſt Beach, and are lost without remedy.'

The Chesil is a single main ridge of pebbles running from just east of Bridport in the west to Portland in the east. Between Abbotsbury and Portland it is separated from the mainland by a lagoon called the Fleet. The overall length of the beach is approximately 29 km, and for only about one-third of this distance is it connected directly with the mainland. In general it becomes larger, both in height and width, towards the Portland end. At Abbotsbury it is 155 m wide and 7 m high, while at Portland (Chesilton) it is over 180 m wide and 13 m high. At the Bridport end the shingle is very fine, scarcely larger than a pea, while at the Portland end, with a regular progression to that point, the prevalent size is 5.0–7.5 cm in diameter. The shingle probably weighs between 50 and 100 million tons, the bulk of it being flint derived from the chalk. There is, however, also some chert from the Greensand, and a few 'foreign' rocks including quartz and black quartzite. Most of these come from the igneous rocks of the South-West Peninsula or from the Budleigh Salterton Pebble Bed.

The Fleet, which is tidal, covers an area of between 3 and 5 square

The magnificent Chesil Beach from the air

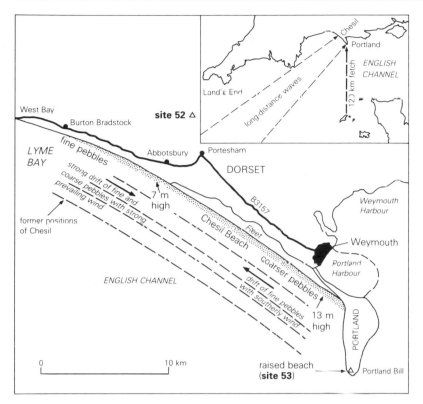

than does almost anywhere else in England. This could help to explain the confinement of the beach between Bridport and Portland, and also its unique size and form in the British Isles. Moreover, Chesil is aligned at approximately right angles to the direction of the long-distance waves, especially in its southern portion.

In spite of the height of the shingle ridge, large storm waves do occasionally break over it. When such rough conditions pile up the sea against the ridge, water passes through the shingle to emerge on the landward side as temporary gullies known as 'caverns' or 'canns'. Sometimes the flow of water is such that the Weymouth to Portland road (and also in the past the railway) is closed by flooding.

The unique size sorting of shingle along the ridge still needs to be explained. The long fetch and prevalent waves from the west and south-west tend to set up a longshore drift towards Portland, and so the majority of the coarse material accumulates at that end. When the winds are more southerly, however, the drift operates in the reverse direction, but the distance travelled by the waves is shorter and so they are less powerful. As a result they are only capable of moving the finer material back towards the Bridport end, thus leaving a sorted ridge.

As the great ice caps melted at the end of the Pleistocene, sea levels started to rise sharply, especially between 13 000 and 6000 years ago: this graph shows the general trend recognised in many parts of the world

kilometres (according to tides), and varies in width from 50 m at the Narrows (SY 650773) to about 1 km at Butterstreet Cove (SY 635799). It is very shallow, varying from 0.3 to 3.0 m (at the Narrows). It was once thought of as either a river valley that ran parallel to the coast but which had been captured laterally by the sea, or as a river which had cut down between the cliffs and the pebble ridge of Chesil itself. However, its origin, as we shall see, must be considered not in isolation but as part of the whole sequence of evolution of Chesil itself. It is in reality the outcome of coastal barrier formation.

An early theory saw the beach as a result of wave action (longshore drift) moving pebbles from the south-west to form a spit extending east from Abbotsbury. This spit then became anchored to Portland at its eastern end. However, this idea is no longer acceptable because the beach

shows no evidence of the curved lateral ridges bending inland which one would expect from such a mechanism. The contrast between the complex form of features such as Hurst Castle Spit, Blakeney Spit, or even Scolt Head Island (site 32) and the relatively simple form of Chesil is very important.

It is much more likely that Chesil is the product of waves developed over a great distance (therefore tending to be larger) combing up vast quantities of material from the floor of Lyme Bay as sea levels rose quickly in postglacial times, between 10 000 and 5 000 years ago. One can envisage a bar slowly 'rolling' landwards with the rising sea level and becoming stranded when no more material was able to be thrown up and over its progressively heightened ridge. A line drawn from Chesil out into the Atlantic shows that this part of the Channel coast receives waves from further away

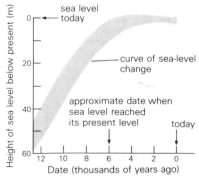

55 A coast of extraordinary landslip and great convulsion

Black Ven from the air, showing a mudflow lobe extending into the sea

The stretch of coast between Axmouth and St Gabriel's, which includes the seaside towns of Lyme Regis and Charmouth, is probably the biggest tract of landslides in Britain. This area is not only of extreme scientific interest but is also one where land-shaping processes have a direct impact on human activities, for the slipping involves the destruction of roads and buildings, and loss of land to the sea.

The geological background is critical to the formation of the landslides. All the slips are associated with the same set of Cretaceous rocks, namely the Chalk (limestone), Upper Greensand (mainly sands) and Gault Clay. These overlie the eroded surface of the Jurassic beds. When it rains the water percolates through the porous Chalk and into the equally porous Upper Greensand. However, it cannot pass into the impermeable Gault Clay, and so it accumulates in the base of the Upper Greensand. These sands, which support the rocks above, are thus liable to be washed away. This in turn causes the overlying sandstones to founder and slide towards the sea, carrying the Chalk (where present) with them.

The sliding is rotational, where the surface of failure (the slip plane) is curved and the rocks tilt backwards as they slide. The top of the old cliff slides downwards and backwards, forming a distinct undercliff, and to compensate the basal layers slide forwards and upwards, pushing up the toe. There have been many reports of offshore reefs or beaches (the toe of the slide) formed at the same time as the downslipping of the cliffs immediately behind.

The slips are associated with periods of abnormally heavy rainfall, which lead to the face of the cliff becoming unstable to a variable depth. Such high rainfalls are known for 1774, 1839 and, very locally, around Pinhay (SY 315913) in 1960–1. The high fall of 1960–1 led to a subsidence of some 6 m in a section of the clifftop at Whitlands (SY 310902), with a rise of some 2 m in a 370 m stretch of beach below (the toe of the slip). However, the most striking feature along the portion of coastline between Culverhole Point and Humble Point is the famous Dowlands Chasm (SY 290895). The great fissure which created it occurred on Christmas night, 1839. This event, which produced 'extraordinary landslip and great convulsion', resulted in a new cliff up to 64 m high being exposed, backing a chasm into which some 8 ha of land had subsided. The length of the chasm was about 800 m, with a breadth from 60 m on the west to 110 m on the east. The foundered

material was estimated to weigh a staggering 8 million tonnes. At the foot of the seaward cliffs a 'reef' of Upper Greensand, nearly 1200 m long, was forced up at sea some 12 m above high-water level; this only survived for a few months before being eroded by the waves. The great chasm, however, shows relatively little alteration from its original condition, except that it has become mantled with vegetation. Its aspect was described thus by the Edwardian topographer, Arthur Norway:

'The reef has been washed away by successive storms; but the cliff ruin stands, though its terrifying aspect is disguised and hidden by the adhesion of the most luxuriant vegetation, which has crept thickly everywhere among the ruin, turning the ravines into bowers of moss and trailing ivy, and strewing primroses and violets so thickly on the slopes that one hardly remembers amid one's admiration of such lavish beauty what murderous agency has been at work. But the memory returns; the sense of awe is dominant; and it is with a feeling of

relief that one climbs up the steep path and sets foot once more upon unshaken ground.'

To the east of Charmouth another major slip is found at Stonebarrow (SY377930). On 14 May 1942 the crest of Stonebarrow broke away and a slice about 500 m from east to west and 18 m from north to south slid down some 15 m into the undercliff of Fairy Dell beneath. (The undercliff is made up of the masses that have slid down from the main cliffs.) The slip has exposed a new cliff of bright yellow Upper Greensand, beneath which two radio-location huts can be seen. These huts were built early in World War II, on top of the original cliff which, when it foundered, carried the huts with it. The displaced slice of cliff on which the huts rest tilts backwards towards the main cliff, giving an excellent illustration of the results of rotational sliding. Stonebarrow is typical of the cliffs in west Dorset, in which a rather remarkable terraced profile is often present, due to the complex nature of the rotational sliding.

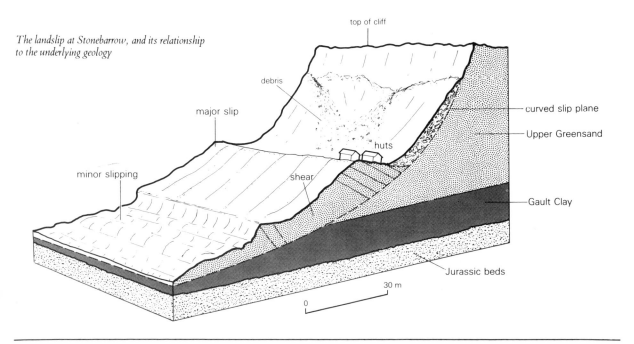

The landslip at Stonebarrow, and its relationship to the underlying geology

top of cliff

debris

major slip

curved slip plane

Upper Greensand

huts

minor slipping

shear

Gault Clay

Jurassic beds

30 m

0

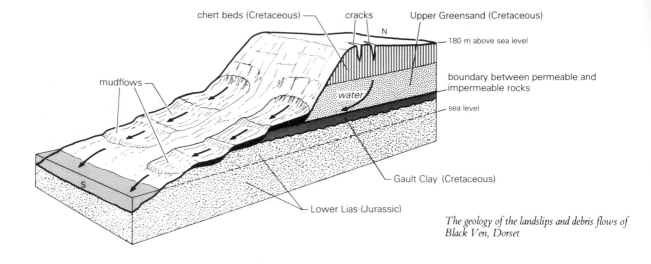

chert beds (Cretaceous) — cracks — Upper Greensand (Cretaceous)

N

180 m above sea level

mudflows

water

boundary between permeable and impermeable rocks

sea level

S

Gault Clay (Cretaceous)

Lower Lias (Jurassic)

The geology of the landslips and debris flows of Black Ven, Dorset

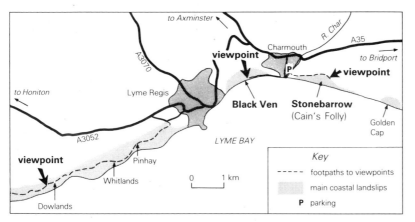

to Axminster

R. Char

A35

Charmouth

viewpoint

to Bridport

A3070

P

viewpoint

to Honiton

Lyme Regis

Black Ven   **Stonebarrow**
(Cain's Folly)

A3052

Golden
Cap

LYME BAY

viewpoint

Pinhay

Whitlands

0   1 km

Dowlands

Key

- - - - footpaths to viewpoints

░░░ main coastal landslips

P parking

Map sheet 193

Viewpoint grid references: Black Ven SY 359932
Stonebarrow SY 377931
Dowlands SY 293896

Facilities etc. in Lyme and Charmouth

The distribution of landslips along the coast near Lyme Regis and Charmouth

There are also other interesting forms of mass movement in the area. These involve a muddy flow of material in a semi-liquid state rather than the wholesale slip of a solid mass as discussed above. As in the case of the slips, alternating bands of clay and limestone and the tendency for water to accumulate at certain layers have provided ideal triggers for the mudflows. Conditions are especially good where the outflow of water from the cliffs is localised. The best mudflow is that at Black Ven, a great amphitheatre-shaped hollow extending from SY 355932 to SY 358932. During periods of wet weather the flows move with noticeable speed, carrying vegetation and rock debris fallen from the cliff above. In many respects the flows exhibit structures found in glaciers, such as crevasses and the like. The source areas of the mudflows are 'corrie-shaped' hollows in which slumping causes the back cliff to retreat. The mud descends through a narrow neck where the main flow is joined by cascades of mud from either side. The accumulated material then pushes forward across the foot of the cliff as a great lobe with a curved front. These flows collect on debris slopes beneath each cliff, eventually creating huge banks. Occasionally the movements are accompanied by the spontaneous combustion of pyrites (an iron mineral), which has led to reports (in 1908 for example) of 'volcanoes' smoking on the Lyme Regis cliffs.

56 Lynton and Lynmouth: a West Country disaster

Devastation caused by the flooding of the River Lyn in the great storm of 15 August 1952

The neck of the South-West Peninsula is particularly susceptible to *heavy* falls of rain, especially when compared to other regions of similar annual rainfall. This is possibly because there are a larger number of convectional storms, triggered off as masses of warm air are forced to rise over the moors. Within the past 50 years the West Country has experienced at least five storms which gave daily falls in excess of 150 mm. The most recent was on 10 July 1968, when 170 mm were registered in the north of the Mendips, causing great changes in

some of the cave systems, and floods in Cheddar (site 47). The outstanding example, however, was the storm of July 1955 when 280 mm was recorded at Martinstown, Dorset. This is the largest amount ever recorded in any one day at a meteorological station in the United Kingdom and it is particularly impressive in an area of low mean annual rainfall. Other great storms include those at Bruton in Somerset (239 mm on 28 June 1917) and at Cannington in Somerset (235 mm on 18 August 1924). However, in terms of the damage caused, the great storm of 15 August

1952 on Exmoor was the most devastating. This memorable summer flood led to massive flows in the Lyn river catchment and extensive landslipping on Exmoor; it killed nine men, sixteen women and nine children; and the resulting damage cost over £9 million to repair.

Despite the records of severe floods in 1607 and 1769, the resort of Lynmouth (SS 723493), where much of the damage occurred, had expanded, and the Lyn River had been obstructed, constricted and neglected in the interests of expanding tourism. The first half of

August that year was wet, with the usual August monthly total falling in the first 14 days of the month. This had largely saturated the soil before the storm of the 15th itself. Then on the evening of the 15th, and the early morning of the 16th as much as 230 mm of rain fell on Longstowe Barrow (SS 710426), at the head of the West Lyn river. About 22730 million litres of water fell over the 100 square kilometres catchment as a whole, giving an areal average of about 140 mm of rain. Because the soil was already full of water from the rain during the preceding fortnight, runoff was extremely rapid, and flood water from the West Lyn converged on the beautiful, but constricted, site of Lynmouth. It has been estimated that with a combined flow of over 520 cubic metres per second during the flood, the two streams (East and West Lyn) had over a period of a few hours a rate of flow almost as great as the record figure for the Thames, which drains an area nearly 100 times greater!

This startling discharge from such a small catchment occurred because much of the catchment has a very steep channel gradient, with some of the contributing valleys falling as much as 425 m in only 6.5 km. Near

the sea, the gradient of the West Lyn reaches an astounding 1:5.6 (c. 20 per cent) and the East Lyn achieves a highly respectable 1:27 (c. 3 per cent). This speeds the flow of water through the system. The effect was multiplied by the temporary ponding up of water by fallen trees and rocks, especially against bridges. This 'physiographic aggravation' combined with the excessive rainfall proved catastrophic. (Part of the steepness of the gradient of the lower Lyn may result from its unusual history involving, possibly, marine capture or glacial diversion, as discussed in site 57.)

The landscape-sculpting power of the storm was considerable. In a single day, the West Lyn moved more than 50000 tonnes of boulders, some of which were over 10 tonnes in weight; a new delta containing some 170000 cubic metres of debris accumulated on the right bank of the East Lyn in Lynmouth, the West Lyn changed its course in Lynmouth, the East Lyn cut across several of its meanders, and a boulder weighing 7.5 tonnes was found in the basement of an hotel. In addition, some major landslides occurred on Exmoor and even today some of the scars still appear fresh where the soil and weathered rock was completely

stripped away.

The landslides are particularly numerous on the south side of the upper Exe valley, north and north-east of Simonsbath, near Warren Farm (SS 800405). Another large concentration occurs in the headwaters of the Hoaroak Water (SS 747423). The movements were of two main types; the majority were slides in which saturated soil and vegetation were stripped off the lubricated surface of the slates beneath, but a few involved flows of mud rather than slides.

The latter type, which one might call a 'debris avalanche', occurred at Warren Farm. It was described thus by an observer who was on the scene shortly after it happened:

'Slaty debris from a deep hole has poured down a slope of fifteen degrees, clearing its path of bracken and mantling the grass with debris. On reaching the small tributary valley at the foot of the slope most of the debris was diverted down that valley to pour into the main valley, but some was carried straight across and up the opposite slope. Such a remarkable uphill flow indicates that the movement was sudden and rapid, and in the nature of a mudflow rather than a slide.'

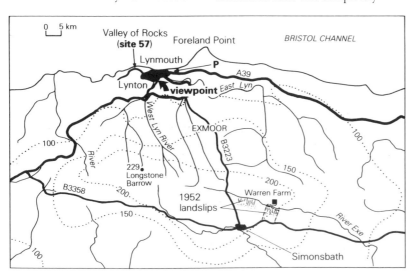

Map sheet 180
Viewpoint grid reference: SS 723493
Cafés etc. in Lynton and Lynmouth

Key

.... rainfall (mm) in storm

- - - - - tracks and footpaths

P   parking

The location of Lynton and Lynmouth, the pattern of rainfall during the great Exmoor storm of 15 August 1952, and the positions of some of the resulting landslips

57　The Valley of Rocks

The spectacular North Devon coast to the west of Lynmouth is tracked by the Somerset and North Devon Coast Path, which takes the rambler within sight of coastal waterfalls, curvaceous cliffs, and some extremely curious valleys. The valleys are mostly dry and over some of their length they run parallel to the coast instead of down towards the sea. One of these is at Hartland Quay (site 59), and another is the oft-visited, and much sketched, Valley of Rocks. This is a curious rock-ribbed valley, now dry, but originally carved by flowing waters of the East Lyn river which has since changed its course. Why should the river change its course? Why does the old route run parallel to the coast? Why is the valley now dry?

One appealing explanation is that the lower stretches of the East Lyn used to flow westwards along the

course of the Valley of Rocks and into the isolated segment of valley where Lee Abbey stands. At this stage, many thousands of years ago, the cliffs would have been well to the north of their present position. Continual battering by the sea

eroded the cliffs and they retreated, eventually dismembering the old course of the East Lyn, first at Lee Abbey and later at Wringcliff Bay and Lynmouth itself. The river, in effect, was 'captured' by the sea, leaving the Valley of Rocks high and

Map sheet 180

Viewpoint grid reference: SS 703497

Car park and café near viewpoint

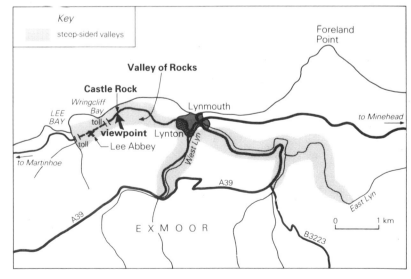

149

The ice margin story

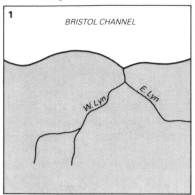

The state of the drainage before the ice sheet in the penultimate glaciation.

The ice sheet arrived, causing ponding-up of the water of East and West Lyn, forming a lake. This water then overflowed westwards forming an overflow channel.

The ice sheet retreated leaving the overflow channel as a dry valley – the present Valley of Rocks.

Two ideas about the formation of the Valley of Rocks

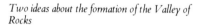

The marine story

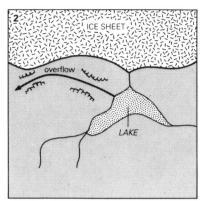

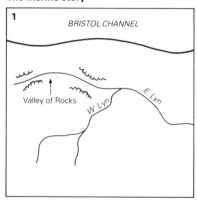

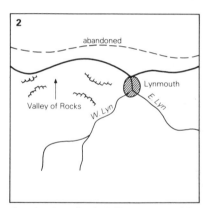

The present-day East Lyn river flowed westwards along the course of the Valley of Rocks. The cliff line was far to the north of the present-day cliff line.

Coastal erosion and marked cliff retreat narrowed the ridge separating the Lyn from the sea. Eventually the cliffs were cut back to the stream bed itself so that the Lyn flowed down the cliff to the sea, at Lynmouth, leaving the Valley of Rocks high and dry.

dry. Support for this idea comes from the fact that the Valley of Rocks, apart from its lack of flowing water, is in many respects similar to the valley of the East Lyn, east of Lynmouth: it is similar in cross section and has the same steep, high valley side slopes. Furthermore, if the slope of the present East Lyn is traced and extended through the Valley of Rocks, the gradient suggests that they were once the same valley.

However, retreat of cliffs along this coast is not especially rapid and the true nature of the valley profiles may have been obscured by large amounts of rock and soil debris in their floors. For these and other reasons, an alternative explanation has been put forward to account for the various peculiar drainage channels along this coast, both at Lynmouth and Hartland Quay. The idea is that the channels were associated with an ice sheet that existed in the area during the penultimate glacial phase of the Ice Age (between about 120000 and 200000 years ago). The ice covered the northern part of the Isle of Lundy, impinged upon the north Devon coast and deposited a boulder clay in the Barnstaple area – the

Fremington till.

The sequence of events might have been as follows. The ice sheet extended southwards across the Bristol Channel and lay with its southern margin along the coast. The Exmoor streams, such as the Lyn, which flowed northwards, found their access to the sea blocked by this ice; as a result they became ponded up. The valleys were flooded until the ponded-back water overflowed the lowest point of the valley side to flow in a direction parallel to the margin of the ice sheet, seeking an

alternative outlet to the open sea. The Valley of Rocks could be one such overflow channel, and those at Clovelly Court (SS 310255), Horne Cross (SS 373236), Hartland Point and St Catherine's Tor (SS 225242) could have a similar origin. As the ice retreated at the end of the glacial phase, the original exit to the sea was freed and the northward-flowing Exmoor streams re-established themselves in their earlier courses. The Valley of Rocks would have been abandoned at this stage, leaving the striking dry valley we see today.

Whichever of these two ideas is correct – and neither seems unreasonable – the Valley of Rocks is also notable for its periglacial landforms and deposits. Under severe tundra conditions, especially in the last glacial phase of the Ice Age some 17000–70000 years ago, frost action was widespread. Freezing of water in joints and weaknesses in the rocks shattered them and prised off angular blocks. Some of these blocks mantle the valley sides as scree slopes. This scree is largely weathered to a dark grey colour and is encrusted with lichens, both features indicating that it is not developing actively now. Some of the shattered debris was, however, removed down into the valley floor by the process of solifluction. This took place during the brief summer interludes when some of the ice melted, saturating the surrounding rock and soil debris. The unstable slurry so formed sludged down the valley sides under gravity. A considerable thickness of such fill is present in the floor of the Valley of Rocks. The removal of soil and rock in the ways just mentioned excavated a group of splendid 'tors' – castellated turrets of rock. Castle Rock, which is exposed right on the top of the valley side, is called a crestal tor, and many valley-side tors are found on the southern slope. This site is a useful reminder that tors are not restricted to granite rocks.

Some of the tors and other landscape features to be found in the Valley of Rocks

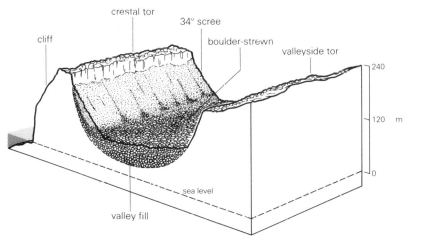

58 Braunton Burrows: a chain of sand hills

The impressive area of sand dunes called Braunton Burrows attracted the attention of Charles Vancouver in his *General view of the agriculture of Devon* (1808, p. 297):

'The burrows ... are composed of a widely-connected and prodigious chain of sand-hills, which have been evidently formed by the force of the westerly winds ... and notwithstanding that the appearance of these hills exhibits the most wild and sterile aspect, and are only appropriated as a rabbit-warren, still there are among them many low and open places, covered with verdure, and formed by the irregularities of the sand-hills, which the wind has cast up, and left in a variety of bold fantastic shapes, forming altogether a striking contrast with their base, which is often seen to rest upon a verdant plain, and giving to the whole an air singularly picturesque, by combining with the view of a desart that of a verdant plain of blooming valley.'

Braunton Burrows form one of the highest and largest dune systems around the coastline of Britain. In places the dunes reach a height of 30 m above the sea, and they extend for over 5 km northwards from the joint estuary of the rivers Taw and Torridge, covering an area of about 800 ha. Compared with other British dunes, they are quite mobile in places. This is partly due to natural causes, but about 40 ha were also reactivated by military activities in World War II. Efforts have been made by the Nature Conservancy Council to plant such active areas with a sand-loving plant called marram grass (*Ammophila arenaria*), which stabilises the sands.

The area is dominated by irregular lines of dunes running approximately north–south and parallel to the coast, but at right angles to the prevailing wind. The highest dunes occur as three or four irregular ranks in the central sector of the burrows. They are separated by low-lying areas called 'slacks'. A blowout occurs when the wind is concentrated and channelled into a certain part of the dune chain, which is hollowed-out as a result. The sand eroded in the process is deposited to form the parabolic dune behind, and the eroded hollow in the original dune forms a low-lying slack. The high groundwater level in these boggy slacks at Braunton has encouraged plant growth, and there is a variety of plant communities, including those dominated by buckshorn plantain (*Plantago coronopus*) and hairy hawkbit

The development of a foredune, and its erosion to give a parabolic dune and a slack

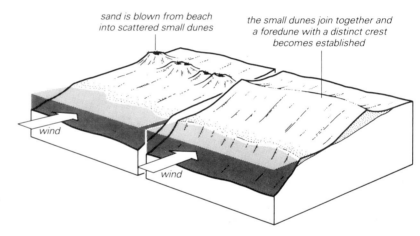

sand is blown from beach into scattered small dunes

the small dunes join together and a foredune with a distinct crest becomes established

wind

wind

a 'blowout' develops at the dune crest and the face steepens as wind erosion continues

a parabolic dune begins to develop

erosion leaves a hollow 'slack' in the foredune

a parabolic free dune is formed

wind

wind

(*Leontodon taraxacoides*). Damp pasture with grasses, sedges and marsh pennyworth (*Hydrocotyle*) is also common, as are communities dominated by rushes or by willows. These various communities provide the 'verdant plains' referred to by Vancouver.

To the north and south of the central area the dunes are less well developed. In the north, partially protected by Saunton Down, they are generally lower, and on the more exposed southern sector often only two ranks of dunes can be identified, the landward consisting of a series of hairpin-shaped dunes, running back alongside the estuary.

It is normally assumed that a series of parallel foredunes on the coast become progressively younger towards the sea and represent a steady outgrowth of the dune system from the land (see site 49). The younger dunes effectively shield the older ones from the sand supply at the beach. However, at Braunton this does not seem to be so, because the inland dunes seem to continue to develop and do not become fully stabilised or cut off. Indeed, the highest dunes tend to be not on the seaward side of the Burrows but further inland, the dunes forming a graded sequence increasing in height in a landward direction. However, landward of the main dunes, covering about half the total area of the burrows, is a region of gentler slopes, with a complex of small dunes interspersed with small hollows liable to flooding. Little sand reaches these sheltered, inactive dunes which are mostly covered by vegetation.

The combination of a good sand supply and a site for sand deposition explains the striking dunes at Braunton Burrows. The coast is fully exposed to the prevailing westerly winds, and at low tide the broad expanse of Saunton Sands provides a plentiful supply of sand. This is brought into the area by the Taw and the Torridge rivers and is spread about the beach by the waves. Saunton Down helps to restrict the

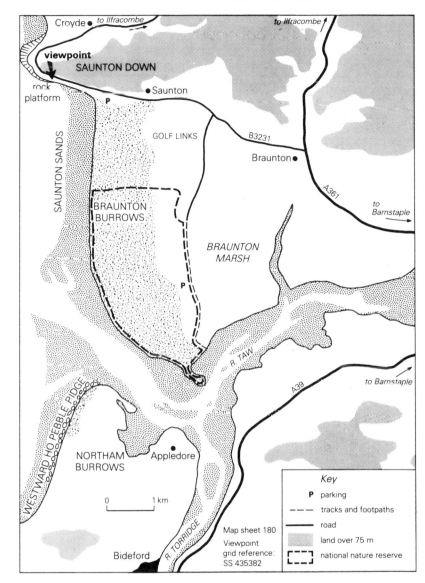

loss of sand northwards. Thus the sand is trapped on Saunton Beach at the mercy of the wind and there is plenty of low ground inland on to which the dunes can move.

Braunton Burrows are now occupied in the centre and south by the military and in the north by a golf club, but a large area has been declared a national nature reserve and is under the protection of the Nature Conservancy Council. The

burrows are best reached along a road, which becomes a rough track, leading off the B3231 (Barnstaple–Croyde) at SS 468374. An excellent aerial view is obtained from the B3231 to the west of Saunton. Limited parking is available off the roadside. On the south side of the estuary there is another sandy area, Northern Burrows, to the seaward of which is the famous pebble ridge of Westward Ho!

59 Hartland Quay and the old course of the Milford Water

Well vegetated dunes and slack at Braunton

E. A. Newell Arber, in his popular book on *The coast scenery of North Devon*, highlighted three particularly curious phenomena: hog's back cliffs, coastal waterfalls, and what he termed 'sea-dissected valleys'. The latter two features are set within magnificent cliff scenery at Hartland Quay and along the coast to the south.

The majority of the rivers and streams of England enter the sea at right angles to the coastline, unless their mouths have been diverted by spits or bars of sand and shingle. In north Devon there are various examples of valleys and streams which, in approaching the shoreline, run parallel to it for some distance before entering the sea. Moreover, many of these valleys have been dissected by cliff erosion, as in the Valley of Rocks (site 57) near Lynton.

To the south of Hartland Quay (SS 233247) there are similar remnants of old courses along the cliff top. The Somerset and North Devon Coast Path follows the cliff line and passes along some of these courses. Starting at the car park near Hartland Quay, follow the path southwards. Soon you will begin to see, just to the right of the path, two flat-floored remnants of a former valley. The remnants have been separated by cliff erosion which, in a few thousand years hence, may have destroyed all trace of these valleys. The path continues across the flat floor of the second remnant, which is separated from the sea by St Catherine's Tor – the former steep valley side truncated by cliff retreat. The Wargery Water, a small stream about 2.5 km long, flows across part of this old valley before cascading down the cliff in a waterfall. It has been unable to cut its channel down to present sea level, and like the old valley it is left 'hanging' from the cliffline.

Just before the path leaves the old valley floor and starts to climb up the steep valley side, stop and look across the spectacular bay framed by high cliffs. On the other side of the bay you will see the waterfall called Speke's Mill Mouth, which cascades from another high, truncated valley, similar to the one you are standing in. In fact, these two features, divided by cliff erosion, are part of the same original valley. The path continues along to Speke's Mill Mouth, offering wonderful views of the coastal scenery on the way.

These channels at Hartland and Speke's Mill Mouth are very different from others along this stretch of coastline: they are broad, flat-bottomed channels eroded into the rock, rather than narrow, V-shaped channels carved through solifluction debris (as at Abbey river (SS 225256) or Welcombe Mouth (SS 213180)) to the south. Therefore, the channels have probably had a different history, but the origin remains obscure. However, the rather exceptional 'trough-like' channels near Hartland are perhaps better explained by the exceptional glacial history of this section of coast. They may well have been formed by ice impinging on the coast and blocking the pre-existing drainage lines, forcing the ponded up water to overflow and carve deep troughs into the bedrock (as discussed in site 57).

The waterfalls are quite interesting in themselves. The best one is Speke's Mill Mouth where the Milford Water forms 'by far the grandest and most imposing waterfall, or, rather, series of falls, on the whole coast'. Its first fall starts at about 50 m above the sea, whence it tumbles 16 m over an enormous smooth slab of shale before plunging into a dark pool. From the pool the Milford Water turns abruptly at right angles and forms a 'gutter fall', within which there are three subsidiary falls. These result from the unequal resistance of the sandstone and shale layers and from the differences in the dip of the beds due to folding. Typically, where the rate of cliff retreat is slow, or where the river is very powerful, or where the rocks are not very resistant, the waterfall is encased in a gully carved into the rocks. The lower part of the Speke's Mill fall illustrates this idea, whereas the upper part of the fall, over a resistant rock structure, is a primitive 'sheer' type.

While in the area it is worth sparing a moment for the superb wave-cut, (or shore) platform developed at the base of the cliffs. It is well exposed at low tide and has formed as a result of the sea attacking the base of the cliffs, which then collapse and recede. In places the platform extends for over 200 m out to sea and it breaks the force of the waves as they approach the coastline. The erosion has taken place across folded rocks, as shown by the patterns of dipping layers, and it is quite easy to reconstruct the original pattern of folds. The folding is so superb as to make this area a geologists' paradise! The platform also shows that harder rock bands stand out as ridges.

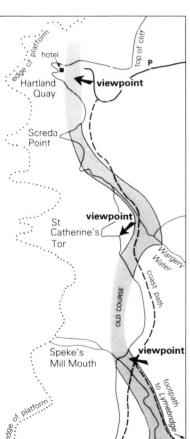

Key

— road

P parking

The old course of the Milford Water near Hartland Quay

Map sheet 190

Viewpoint grid references: Hartland Quay SS 223247
St Catherine's SS 224244
Speke's Mill SS 225236

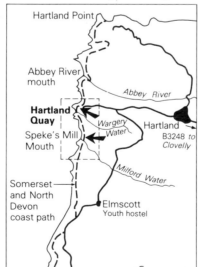

60   The Dartmoor tors: Cyclopean masonry

Hay Tor – the most visited of the Dartmoor tors

Some 300 million years ago the Dartmoor granite, generally grey in colour and often coarse in grain, was forced up as molten rock from within the Earth, into Devonian slates and sandstones. This molten rock gradually solidified and the forces of erosion stripped away the overlying rocks to reveal the granite as a great dome (a batholith) rising up above the surrounding lowlands. This granite, more resistant than the surrounding rocks, gives rise to the highest land in the West Country, exceeding 620 m at High Willhays.

Although resistant, the granite is not immune from the effects of weathering and erosion. It contains many joints and other lines of weakness along which these activities tend to be concentrated. Some of the granite is much broken up by joints, other parts less so. These joints, together with their variable spacing, give rise to one of the most striking features of Dartmoor – the tor.

As David Linton, an eminent scientist, wrote:

'They rise as conspicuous and often fantastic features from the long swelling skylines of the moor, and dominate its lonely spaces to an extent that seems out of proportion to their size. Approach one of them more closely and the shape that seemed large and sinister when silhouetted against the sunset sky is revealed as a bare rock mass, surmounted and surrounded by blocks and boulders, rarely will the whole thing be more than a score or so feet high. But if on closer examination the tor loses something of its grandeur, it loses nothing of its strangeness.'

Indeed, climb the tors themselves and on the flatter-topped varieties find the shallow rock basins once supposed to be druid sacrificial altars. Examples of such basins, about a metre across and a metre deep, are well displayed on many of the tors, and their existence is recorded on the OS 1:50000 sheets. These are natural features, somehow formed by weathering.

The origin of the 'cyclopean masonry' that makes up the tors has been the subject of major debates and conjectures. Early theories suggested that they were man-made monuments, relict sea stacks, or residuals of rock chiselled out by wind erosion, but more recent ideas involve weathering under conditions that no longer exist. David Linton believed that they developed in two stages. The first saw intensive rotting of the granite under warm conditions to give rounded boulders set in a mass of decomposed rock known as growan; the second involved the stripping of the growan to expose the rounded boulders, which look like a 'great heap of piled woolsacks', and which make up the tors.

Alternatively, some people think the tors were excavated by severe frost weathering of the solid granite in the Ice Age, and that the bulk of the deeply decomposed rock forming the lower-lying areas is the product of the baking of rocks (metamorphism) which formed the well known china

Deep weathering followed by stripping

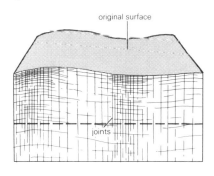

Two ideas about the formation of the Dartmoor tors

Rock baking (metamorphism) followed by frost action

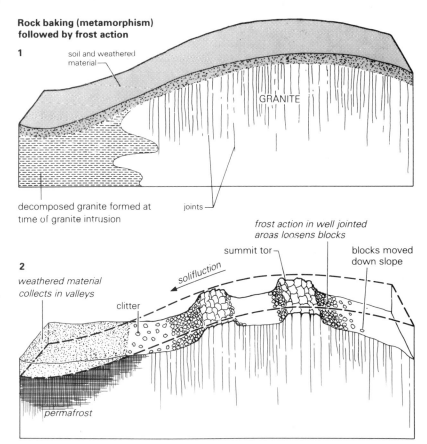

clay (kaolin) deposits. Areas of close jointing would be severely attacked by the frost, whereas the rock would be little affected where the jointing was less dense. Stripping of the frost-shattered debris would leave the coarsely jointed rock masses as upstanding tors.

One mechanism that could have caused the stripping is solifluction; periodic sludging of debris down slope under arctic conditions. This would also account for the great lines of angular granite blocks (clitter) which frequently radiate from the buttresses and tors of the moorland summits. This debris often assumes a striped pattern, similar to that forming today in many periglacial areas. The view from Cox Tor (SX 531762) across to Staple Tor (SX 542760) shows these stripes magnificently.

The tors thus stand as residual rock piles. They survive as monuments in a landscape held in a delicate balance between the rate of formation of weathered debris on the one hand, and the rate of its removal on the other. There are probably several different ways in which the rock can be weathered and the debris removed; all resulting in a similar landform – the tor.

61   River capture and the Lydford Gorge

When one river is able to cut back more effectively than one of its neighbours, it may succeed in 'capturing' the headwaters of the other stream. It thereby increases its own discharge and power, reducing the other stream to a mere trickle within an over-sized valley – a misfit (see site 43).

River capture (or 'river piracy') appears to have been widespread in the West Country. The watershed between rivers flowing into the Bristol Channel and those making their way to the English Channel lies mainly on the northern side of the peninsula. But breaches in this watershed indicate that over time there have been some 'contests' between northward and southward flowing drainage that have given rise to river capture. The Hayle, the Camel and the Torridge rivers show evidence of such competition in course.

Undoubtedly one of the most impressive examples of river capture and its associated landscape features is the beautiful but sometimes sinister Lydford Gorge. This deep, narrow valley is now owned by the National Trust and is best entered at its northern end, near Lydford (SX 508845). A good path and innumerable steps have been constructed along its length. The long route takes you on a round tour of the gorge; through garlic-scented woodlands, across narrow chasms spanned by wooden bridges, and down to the base of the streaming White Lady Falls. It then meanders peacefully beside the Lyd through the base of the gorge – an enchanting stretch of pools, potholes and water-worn rock faces. In one or two places the path clings to the edge of the rocky ledges, just above the river. Eventually it brings you to the crashing Devil's Cauldron in the upper reaches of the gorge. Approaching along a narrow, be-ferned side track, oozing moisture and dampness, you are led right into the bowels of one of the many giant potholes. Perched on planks less than 1 m above the water level, you are surrounded by the clamour of the foaming torrents. It is a fitting end to

Lydford Gorge – the approach to the Devil's Cauldron

a magical walk. The return is fairly easily accomplished along the short route that leads directly back to the main entrance.

The gorge has long been a celebrated part of the local landscape, and its attractions were summed up thus by the Edwardian writer, Arthur Norway:

'There are few scenes in Devonshire so remarkable as this. One comes along an ordinary country lane, for here the soil is not waste, but cultivated, and dropping down a little hill the bridge lies straight in front, looking in no wise different from any ordinary piece of stonework that carries a country road across a rustic brook, until one stands upon it. Then it is seen that the valley below is not only exceptionally deep, but that in the

bottom of its richly wooded slope the stream has cut down straight and sheer a deep black gorge through the solid rocks, where it thunders on in shadow, far beyond the reach of sunlight, a gloomy torrent in whose sound the steady singing of the moorland streams is deepened into a harsh threatening roar ... The gorge is deep enough to be majestic.' (This view can be seen from the bridge just before the entrance to the National Trust car park.) Some visitors have, it is said, been rather less enthusiastic, and among these was William Gilpin, who described it as 'a mere garden scene' and who regarded it as 'a spout rather than a cascade'.

The gorge at Lydford, an impressive 46 m cleft walled by Upper Devonian slates, results from the diversion of the Burn to a shorter

course with a steeper fall, brought about by the breaching of the sides of its original valley by the River Lyd. Previously the Burn flowed south from Bridestowe and Sourton Common, past Wastor Farm (SX 497827) and along the deep wide valley now drained by the relatively insignificant remnant of the River Burn, a tributary of the Tavy. The capturing River Lyd had a steeper course and a more powerful erosive ability, draining westwards to the River Tamar near Lifton (SX 375840). Once it had eroded far enough back, the Lyd tapped the waters of the southward-flowing Burn. Thus a major set of waterfalls was created as the Burn, flowing at a higher level, tumbled into the course of the Lyd. Gradually, using its recently captured energy, the Lyd started to deepen its new upper valley, so as to even out the profile of the river. In the process the waterfall retreated and diminished and the Lydford Gorge was carved out.

The point of capture is marked by a characteristic elbow of capture, and the gorge extends upstream from here. Part of the original Burn Valley south of the gorge is now dry (a dry gap) and the Burn starts further south. A small misfit stream drains northwards from the dry gap into the River Lyd. The tributaries and the misfit stream have been unable to keep pace with the Lyd's incision and consequently they cascade from their original high valley down into the gorge below as sheets of falling water. The White Lady Fall, tumbling from a 'hanging valley' into the gorge, is an example.

The erosive power of the Lyd in its gorge tract is illustrated by the development of many large potholes in its bed and sides. These are formed by a combination of the sheer pressure of the water and by the powerful eddies picking up stones and battering them against the rock surfaces. The most impressive example of a modern pothole is the Devil's Cauldron.

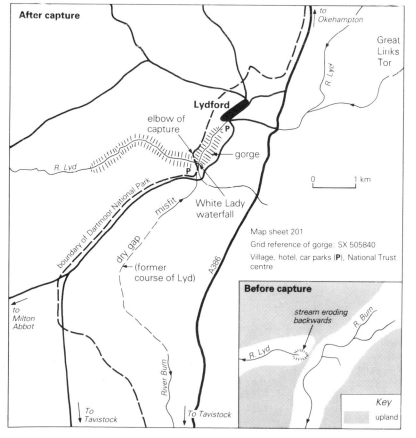

After capture

to Okehampton

Great Links Tor

R. Lyd

Lydford

elbow of capture

P

gorge

R. Lyd

P

0   1 km

boundary of Dartmoor National Park

misfit

White Lady waterfall

dry gap

(former course of Lyd)

A386

to Milton Abbot

River Burn

To Tavistock

To Tavistock

Map sheet 201

Grid reference of gorge: SX 505840

Village, hotel, car parks (P), National Trust centre

Before capture

stream eroding backwards

R. Burn

R. Lyd

Key

upland

62   Hallsands: the sea's revenge

In the 19th century, near the southernmost point of Devon, there was a busy little fishing port of Hallsands, built on a small platform between sea and cliff at the southern end of Start Bay. Today it is almost deserted and only a few of the original cottages remain. On a grey day it is a forlorn site, for the ruins of the fishermen's houses are still standing, as they have since the village was ravaged by a severe easterly storm in 1917. Yet it was not merely 'an act of God' that brought about this unhappy fate. Man himself played a significant role.

At the turn of the century when Britannia ruled the waves with her imperial navy, extensions were made to Devonport Naval Dockyard. In order to make concrete, a contractor, Sir John Jackson, was licensed by the Board of Trade to 'dredge and carry away sand, shingle and other materials from that part of the sea bed, between the high and low water marks at Start Bay and opposite Hallsands and Beeson Sands (Beesands)'. Much to the concern of the villagers, he was soon removing 1600 tonnes of shingle *each day* from the beach. Upwards of half a million tonnes were removed, lowering the beach by 5 m in places. The original condition of the contract was ignored, namely that dredging should not constitute a danger to the coast or expose the land above high water mark to encroachment by the sea.

It had been anticipated that nature would replenish the shingle that man had taken away and that the beach would be re-formed, thereby continuing to protect the village on its precarious platform between sea and cliff. A good beach is a wonderful way to dissipate the energy of storm waves. However, nature was not bounteous and the beach was not replenished. The extent of the depletion can be gauged by considering Wilson's Rock, a small stack just offshore. Old surveys and photographs show that this rock was once completely covered with shingle. Today it is not.

The reason for this lack of replenishment, which exposed Hallsands to the full fury of the waves in 1917, was that the beach was of great antiquity and formed from a once-and-for-all supply. The shingle was dumped on the beach possibly as long as 6000 years ago as sea level rose following the melting of the ice sheets. The rising sea combed up shingle from the sea floor of Start Bay. Since sea level became stable, around 6000 years ago, this

Hallsands village

before dredging

1903

1975

7 m

sea level

shingle

Wilson's Rock

The change in the beach profile at Hallsands as a result of the dredging of shingle

Ruined fishermen's cottages loom up out of the mist at Hallsands .

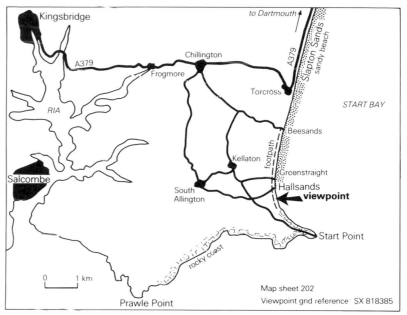

combing process has ceased, so that when man removed the shingle there was no source to replace it, leaving Hallsands at the mercy of the waves.

Today Torcross, a little further to the north, suffers from time to time in much the same way as Hallsands. This may be a long-term result of the shingle dredging over 80 years ago. Torcross is easily reached along the A379 from Kingsbridge. For Hallsands, turn off at Chillington and follow the signposted lanes. Alternatively, walk around the bay from Torcross to Hallsands.

63   Roughtor: nature's sculpture

Although tors are synonymous with Dartmoor, they are also richly developed elsewhere in the South-West, notably on parts of Bodmin Moor which was intruded as magma into overlying sedimentary rocks some 270 million years ago. As an Elizabethan traveller, Norden, remarked:

'The Inlands mountayns are so crowned with mightie rocks, as he that passeth throwgh the Countrye beholdinge some of theis Rockes afar off, may suppose them to be greatye Cyties planted on the hills, wherein *prima facie* ther appeareth the resemblance of towres, howses, chimnies and such like, Crags and

Rockes vncouered (as may be thowghte) and left bare-headed at the Vniuersal inundation, whose force searching the verie foundations of the yelding earth, carried with violence heapes thereof togeather, making mountaynes of valleys, and of valleys loftie hills.'

Bodmin Moor, which hardly rises above 300 m, forms the highest land in Cornwall. It has an origin comparable to Dartmoor and is intermediate in size and height between its Devonian neighbour and the smaller granite masses, possibly the tips of a large intrusion, such as St Austell Moor, Carn Menellis, Land's end and the Isles of Scilly. Like

Dartmoor and St Austell Moor its margin is rimmed with zones of kaolin which were probably produced during the cooling and consolidation of the granite.

Roughtor (National Trust land; SX 145808) is very aptly named in view of the great spreads of boulders (clitter) one has to scramble over to reach the summit tors. It is one of the most striking of the Bodmin Tors and it provides a fitting monument to men of the 43rd (Wessex) Division who were killed in World War II. Standing at over 400 m, Roughtor provides exciting views over the moor, to the coast, to the china clay works and their artificial mountains, and across the many prehistoric stone circles on its slopes.

The summit of Roughtor is crowned by several groups of tors, each looking like a heap of woolsacks stacked one on the other. The tors rise up from clitter-covered slopes that radiate for as much as 1 km from the summit. The large, angular blocks of clitter are clearly visible beneath later river sediments and peat in the banks of the stream just to the south of the car park (SX 138818). On the southwestern side of the tors the clitter is deposited in the form of great lobes. Much of it is now covered in lichens and embedded in turf and is plainly a fossil feature. There can be little doubt that solifluction and frost shattering during the Ice Age have played a major role in shaping these tors and the adjacent slopes. Further discussion of some of the main ideas on tor formation can be found in our description of site 60.

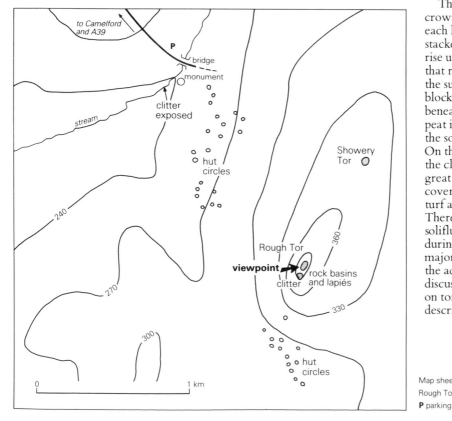

Map sheet 200
Rough Tor grid reference: SX 145807
P parking

Roughtor from the north-west, looking across hut circles

Today, some chemical weathering of the granite is taking place. The southwestern summit tor, for example, is only partly covered by lichen, and it appears white in colour rather than grey, where material is flaking off to expose fresh rock surfaces. In addition, some very impressive minor weathering features have developed on the surface of the tors. The edges of the summits, for example, look like a pie-crust or the toes of elephants, for they have a curious fluted pattern (called lapiés), with small ridges and troughs (about 0.25 m across). Similar features are sometimes encountered on the surface of limestone that has been attacked by solution. In contrast, the flat tops of the tors often have rock basins developed in them. The basins are shaped like frying pans (20–60 cm in diameter and 20 cm deep), and have channels running from them. The rock in their bases appears to be fresh and it is likely that they are still forming today. Similar basins are found on the Dartmoor tors.

64 Fowey ria and the drowned coastline of Cornwall

The sea perhaps has a greater effect on the scenery of the south-west peninsula than on any other part of England, on account of the great indentation of the coastline. Long fingers of sea push a considerable distance inland, providing harbours with sheltered waters as at Falmouth, Dartmouth and Fowey. These inlets are flooded river valleys, to which the name 'ria' is given. Although rias are best developed on the southern shore of the south-west peninsula, they also occur in the north. Thus, the Severn Estuary is in part a great ria, and the Taw–Torridge Estuary at Barnstaple seems to have eroded a channel down to about 46 m below present sea level.

Fowey ria has been a safe haven for generations of seafarers, and all the greatest sea traditions of Cornwall are locked up in it. There was a time when this Cornish harbour led the kingdom in matters of seamanship. When Edward III collected a fleet for the siege of Calais, Fowey contributed 47 ships and 770 men. Yarmouth sent 43, Dartmouth 32, and London a paltry 25.

The rias have formed as a consequence of changes in the relative levels of land and sea since the beginning of the Ice Age. In the glacial phases three times as much water was trapped in the form of ice over the Poles and large parts of

Scandinavia and North America than is the case today. Sea levels fell throughout the world by 100–150 m exposing large areas of the continental shelves to the action of rivers and the elements. Thus, for example, the water drained from the North Sea and the Irish Sea, leaving them as dry land. This meant that rivers had to adjust to a new base-level of erosion, and this they achieved by cutting down their beds towards the new low sea level of glacial times, eroding deep valleys.

These same deep valleys were in turn flooded by the postglacial rise in sea level as the ice sheets melted. At the end of the last glacial the sea level rose remarkably quickly, around

Fowey Ria – a drowned river valley with a notable maritime history

Original high sea level

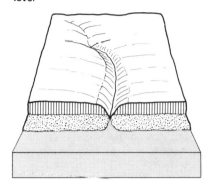

The sea level falls, and the river cuts down (during a glacial period)

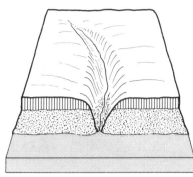

The sea level rises again and the valley is flooded

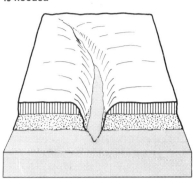

The formation of a ria like those of Fowey and Solva (site 26) by sea-level changes

At Fowey and Solva (site 26), the floor is filling to a new level

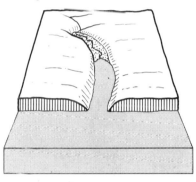

At Gribin (site 26) – the ria is already infilled with sediment

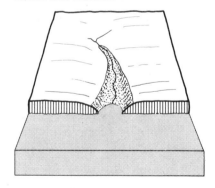

100 m between 17000 and 6000 years ago. The effects of this submergence (called 'the Flandrian transgression') on Chesil Beach (site 54) and the Pembroke coast of Wales (site 26) have already been described. The transgression also flooded low-lying woodland to give submerged forests that at low tide can be found widely around the Devon and Cornwall coastline.

Fowey, one of the best examples of a ria in Britain, is only 300 m wide at its mouth, but it extends inland for about 8 km with a width of 200–300 m. It is enclosed by a plateau surface at around 100 m into which various tributary valleys are cut, creating creeks ('lateral rias') such as Penpoll Creek, Mixtow Pill and Pont Pill.

The creeks and the upper reaches of the Fowey itself show evidence of silting up. The main channel meanders from side to side of the estuary between mudflats which are exposed at low tide. This choking by silt results from natural processes of sedimentation in postglacial times, though in some parts of the South-West the rate of silting (termed 'alluviation') has been accelerated by the activities of man. The tin miners and the china-clay workers have often added large quantities of sediment to the streams, and this is deposited where the streams reach the calm waters of the rias.

A good locality at which to see the Fowey river and its lateral creeks is

Map sheet 200
Viewpoint grid reference: SX 124547

the village of Golant (SX 123547). The small quay on the ria side of the railway track gives the best view and can be reached via a level-crossing at the south end of the village.

65 The Loe Bar and Pool: Cornwall's largest lake

South-east of Porthleven, on the shore of Mount's Bay, is the largest lake in Cornwall, the Loe, which is separated from the sea by a large bar of shingle (SW 643240). The shingle is composed dominantly (90 per cent) of flints, a material that does not outcrop locally, neighbouring cliffs being composed of schist (a metamorphic rock) and quartz.

The Loe itself is a freshwater body over 2 km in length, about 250 m wide. It has two main lateral creeks, Carminowe Creek on the east, and Penrose Creek on the west. The lake shore is between 6 and 7 km long, and the surface area of the water is about 60 ha. The depth of water varies, but reaches 12 m. The great bulk of the water comes from the River Cober.

The extraordinary nature of this lake and bar was noted at length by most of the early topographers and natural historians. Leland, writing in the reign of Henry VIII, reported that the river was prone to break through the bar, 'And one in 3 or 4 Yeres, what by the wait of the fresch Water and the Rage of the Se, it brekith out, and then the Fresch and salt Water metyng makith a wonderful Noise'. Likewise, Borlase wrote of the 'Lo Pool' in 1758, 'The lake is about two miles long, and a furlong wide, formed by a bar of pebbles, ſand and ſhingle, forced up againſt the mouth of this Creek by the South Weſt winds; the valley betwixt high lands on each ſide

giving vent to, the thereby increaſing the force and velocity of the winds from this quarter ... This lake is remarkable for an excellent and peculiar trout'.

In past times the pool has been subject to flooding, though the construction of an overflow at its north end in the middle of the 19th century improved matters. Before this it was sometimes necessary to cut the bar to allow the fresh water to escape, and Borlase reports that, in order to be allowed to do this, the Mayor of Helston had to send a leather purse with three halfpence in it to the Lord of the Manor of Penrose. The bar was last cut in the winter of 1867–8. On occasions, as Leland and others make clear, the breaching of the shingle bar would take place naturally. However, later reports give no indication of an outlet being formed by natural forces, and there is little doubt that this ceased at some time between 1600 and 1800. Surveys show that the bar is

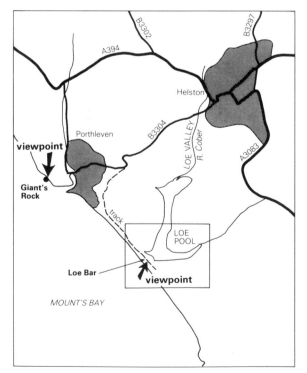

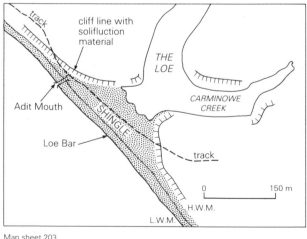

Map sheet 203
Viewpoint grid reference: Loe Bar SW 643243
Giant's Rock SW 623257

currently increasing in height and moving landwards.

One theory for the formation of the bar and pool is that two spits developed, one each side of the mouth of the creek. For a while a channel was maintained between them by the flow of water from the Cober and by tidal scour. Then the channel was closed completely, either by gradual processes, or by a tidal wave or sudden storms, such as silted up the entrance to the Rother at New Romsey in the 13th century. Once the bar had been formed, storms would pile more and more shingle up on to it until the wash of the waves could reach no higher except in very severe storms. Part of the bar is now covered by vegetation, suggesting it is fairly stable. Water from the pool trapped behind seeps out to sea through the shingle ridge.

However, it is also possible that the shingle ridge originated as an offshore bar and was driven onshore as the sea level rose at the end of the Ice Age. Chesil Beach (site 54) is the finest example of this process, and many of the major shingle features on the south coast probably formed in the same way. The material in the ridge would, therefore, be derived from offshore. This could explain the presence of flints in the bar. Local material, such as might be driven by longshore drift from the nearby cliffs, seems to form little of the shingle. Another bar which encloses a lagoon and may have a similar origin is at Slapton Ley (SX 825430) between Dartmouth and Start Point in south Devon.

Another interesting feature in the area, less than a kilometre to the east of the harbour at Porthleven (SW 623257), is the giant boulder lying on the shore. It is called the Giant's Rock and is one of the largest 'erratic' blocks in the South-West. It weighs 50 tonnes and is of a rock called 'gneiss'. It rests in a hollow on the heavily grooved shore platform, and associated with it are four smaller boulders of 'quartzite' (so called because it consists mainly of

quartz). There has been considerable controversy as to its origin. On the one hand some people believe that it was deposited as a true erratic block when, quite early on in the Ice Age, the glaciers may have entered parts of the West Country, while on the other it is sometimes argued that the boulder was dropped by one of the many icebergs that would have appeared in the Channel during the glacial era.

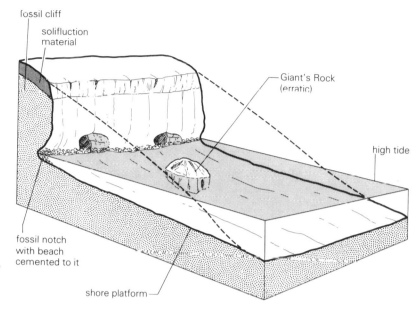

The Giant's Rock at Porthleven, in relation to the shore platform, fossil cliff and beach, and solifluction material

The Giant's Rock is a 50 tonne gneiss erratic dumped at the coast by ice

Glossary

anticline   An upward, arch-like fold in a set of rocks.

arête   A precipitous ridge left between two adjoining corries.

basalt   A fine-textured, dark coloured igneous rock formed when molten lava extruded on to the surface of the Earth from a fissure or volcano solidifies.

batholith   A large mass of igneous rock pushed (intruded) into the Earth's crust. Subsequent erosion of the overlying rocks may reveal the top of the igneous mass at the surface.

blockstream/field   An accumulation of boulders in the bottom of a valley (blockstream) or on a hillslope (blockfield). Typically, the boulders are made of tough **silcrete**, and are the eroded remnants of a much larger sheet of **silcrete**.

blowout   Disturbance, often by man or animals, of an ancient sand dune may result in renewed removal of sand from the area, leaving a hollow – a blowout – in the former dune. Blowouts often form when stabilising plant cover on the dune is destroyed.

boulder clay   The mixture of gravels, sands, silts and clays dumped from a glacier or ice sheet as it melts and retreats. The boulder clay often forms a hummocky 'cloak' on the landscape of areas once covered by ice.

catchment   The area of land that drains into a particular stream or river network.

chert   A tough, glassy-looking material made up of minute crystals of silica. It often occurs as nodules in **limestone** rocks, and is tan/black in colour.

clints   The blocks of rock that make up a bare **limestone** surface.

clitter   Angular, rock debris littering the hillsides around the base of tors. It is formed mostly under very cold conditions by frost shattering of the exposed rocks and sludging of the debris downslope (**solifluxion**).

col   A smooth saddle of land between two hills.

convectional storm   Strong heating of the land surface in summer results in the formation of towering cumulus clouds and intense short downpours of rainfall, possibly with thunder and lightning.

coombe   A short valley, usually lacking water (a dry valley) carved into the steep scarp face of a limestone ridge. Typical examples are found on the chalk scarps of the North and South Downs.

corrie (cwm, cirque)   A large, armchair-shaped hollow, gouged out of the side of a valley or rockface in a highland area. It is hollowed out by a combination of ice accumulation and movement, and frost shattering of the rocks.

dip slope   The gentle backslope of an escarpment, where the land surface is usually sloping at an angle similar to that of the underlying tilted rocks.

dolerite   A medium-textured, dark coloured igneous rock, formed by the cooling of molten magma within the upper layers of the crust. It often forms **dykes** and **sills**.

dyke   A sheet of igneous rock intruded into the Earth's crust, whose form bears no relation to the structure of the existing rocks around it.

erratic   A large pebble or boulder that has been transported, usually by ice, some distance away from its source into an area of different rocks – an immigrant boulder.

esker   A long winding ridge of gravel and sand dumped by a subglacial stream.

Flandrian transgression   The rise in sea level following the widespread melting and retreat of ice sheets at the end of the Ice Age. It commenced around 15 000 years ago, and the sea finally reached present level about 6000 years ago.

glacials   Cold periods during an ice age when ice sheets expand to cover much larger areas than at present.

gneiss   Tough rock formed from alteration of pre-existing rock under conditions of very high temperature and pressure. Such conditions are associated with the intrusion of a mass of magma into the Earth's crust.

granite A coarse-textured igneous rock formed as molten magma pushed into the Earth's crust cools slowly. **Batholiths** are usually composed of granite.

grikes Clefts separating blocks of rock (**clints**) that make up a bare **limestone** rock surface. The grikes are formed by solution of the **limestone** along weaknesses.

gritstone A sedimentary rock made of coarse sands or gravels cemented together.

head An illsorted mass of soil and angular rock fragments formed by frost shattering that has sludged from the hillsides into the valleys by the process of **solifluxion**.

Holocene Postglacial time – the present warm period that commenced at the end of the Ice Age, about 11 000 years ago.

interglacials Relatively short, warm periods during an ice age when climates were similar to the present and ice sheets contracted.

lacustrine This refers to lakes, and so a lacustrine deposit is the sediment that accumulates in the floor of a lake. Such sediments are often layered horizontally.

limestone Rocks made up of carbonate, mostly calcium carbonate. Generally these are formed beneath the sea, and the carbonate may either come from the skeletal remains of plants and animals (as in corals and chalk) or it can be precipitated directly (as in **oolitic limestone**).

limestone pavement A bare, smooth surface of limestone, either flat or sloping, formed mainly by the scouring action of moving ice. Solution along weak joints may divide the pavement into **clints** and **grikes**.

loam Soil that contains roughly equal amounts of sand (2–0.06 mm), silt (0.06–0.004 mm) and clay-sized (less than 0.004 mm) material.

long profile The profile of a river's course from its mouth to the sea. It is often shown as a graph of the height of the stream bed above sea level against distance from the sea.

longshore drift The movement of beach material along the shore by the action of waves.

mass movement The movement of soil and loose rock debris down slope under the force of gravity. There are several different types of mass movement including rock falls, landslides, mudflows and soil creep.

meander A natural horseshoe-like bend in the course of a river or stream.

meander core Where the meanders are very sinuous, turning almost back upon themselves, the small area of higher land in the centre of the meander loop is termed the meander core.

moraine A poorly sorted mass of material carried on, within or at the base of ice, that is dumped, as the ice melts, in the form of a ridge or mound. Moraines often mark the former edges and tips of glaciers and ice sheets.

oolitic limestone A type of pale-coloured **limestone** made up of small rounded spheres of calcium carbonate precipitated from sea water.

overflow channel A channel carved into the land surface by waters overflowing from a lake or flooded area. Such channels often form when meltwater from glaciers or ice sheets is ponded up until it overflows.

parabolic dune Crescent-shaped sand dune formed when a partially vegetated ridge of sand is disturbed. The wind hollows sand out of the disturbed area, leaving a **blowout**, and the sand is blown down wind and deposited as a parabolic dune, with its horns pointing in the direction from which it has come.

periglacial An area which experiences very cold conditions but is not covered by glacier ice. **Permafrost** is common in these areas, as is the sludging of material downslope by **solifluxion**.

permafrost Permanently frozen ground that forms under very cold conditions, typically in areas near to ice sheets. In extreme cases it may extend to depths of 500 m below the ground surface. The top layers often thaw in the summer months.

pH A measure of the acidity or alkalinity of a soil. Higher pH values (from 8 to 14) indicate alkaline conditions, whereas lower pH values (from 1 to 6) indicate acid conditions.

pingo A mound formed by the growth of a lens of ice underground forcing the land upwards. These are usually formed under arctic conditions.

Pleistocene Commonly known as the Ice Age – the period from about 2 million years ago until 11 000 years ago, characterised by alternating cold and warm episodes and associated growth and decay of ice sheets and glaciers.

radiocarbon dating A way of obtaining the age of materials rich in carbon by measuring the amount of radioactive carbon present within them. It is based upon the idea that the radioactive element decays at a constant known rate over time.

ria A river valley that has been drowned in its lower reaches by the sea. This can result either from the sea level rising or from the land sinking. Rias often form good, sheltered harbours.

rill A small channel formed by water running down a hillslope. Unlike a stream, water only flows in a rill during and after heavy rainfall.

roddens Pale-coloured silty ridges that are the remains of banks of creeks and rivers that once drained the fenlands.

rotational slide/slip A type of landslide where the mass fails along a curved surface beneath the ground, and individual blocks of rock tilt or rotate backwards as the whole mass slides forwards.

sandstone A sedimentary rock composed of particles ranging in size from 1/16 mm to 2 mm and held together by a cementing agent. The particles are derived from the erosion of other rocks and may be accumulated by wind or water action.

scarp A steep, short slope that rises up from a low-lying vale or plain, and that is backed by a gentle **dip slope**. Scarp and **dip slope** together form an asymmetrical ridge, often in relatively resistant rocks.

shale A soft sedimentary rock derived from the erosion of other rocks. It is composed of very small (less than 0.004 mm) clay particles that accumulate in water. The parallel alignment of the clay particles makes it easy to split the rock.

silcrete A tough dense material composed almost entirely of the mineral silica. It is often found as layers on and within the chalk downs of southern England. Disintegration of these layers gives rise to the silcrete boulders found in blockstreams.

sill A sheet of igneous rock formed as molten magma cooled within the crust. The magma was pushed between two layers of rock, and so the sill conforms with the structure of the surrounding rocks.

siltstone A soft sedimentary rock formed by the accumulation of silt particles (0.06–0.004 mm) under water. Particles are derived from the erosion of other rocks.

solifluxion Thawing of the top layers of frozen ground (**permafrost**) will create a mass of soggy soil and rock debris. Solifluxion is when this mass flows down slope under the influence of gravity, and it can take place on very gentle slopes.

solution Rainwater, which is always slightly acid, easily dissolves **limestone** (calcium carbonate) as it passes through the rock. This process of removal of the **limestone** is termed solution.

stack An isolated pinnacle of rock rising out of the sea that has been separated from the main coastline by erosion of the intervening rocks.

striations Small grooves on a rock surface usually caused by stones and boulders in the base of a glacier being dragged across the rock.

surface of failure The underground plane of weakness along which a landslide takes place.

swallow hole A hole or hollow in the landscape caused by solution and/or collapse of the underlying **limestone**. Surface streams may disappear underground down such a hole.

syncline A downward, trench-like fold in a set of rocks.

tombolo A bar of sand that joins an island to another island or to the mainland.

tufa Calcium carbonate (**limestone**) precipitated chemically from fresh water containing abundant calcium carbonate in solution. It is common around waterfalls and springs in **limestone** areas, and also forms stalactites and stalagmites in caves and caverns.

watershed The boundary between two or more independent river systems.

It is usually the area of highest ground. Rain falling on each side of the watershed drains into different river systems.

water table The level beneath the ground below which the rocks are saturated with water (groundwater) and above which they remain unsaturated. The water table fluctuates during the year due to changing weather conditions.

weathering The physical breakdown and/or chemical decomposition of rocks at the Earth's surface as a result of the action of ice, water, wind, temperature fluctuations, plants and bacteria.

Further reading

For general background reading on the study of landforms – geomorphology –
there is a wide range of useful textbooks available that carry British examples.
These include:

B. W. Sparks   *Geomorphology* (Longman)
B. W. Sparks   *Rocks and relief* (Longman)
G. H. Dury   *The face of the Earth* (Pelican)
R. J. Rice   *Fundamentals of geomorphology* (Longman)
R. J. Small   *Landforms* (Cambridge University Press)

For a study of coastal features see J. A. Steers *The Coastline of England and Wales*
(Cambridge University Press), for glacial and periglacial features see B. W.
Sparks & R. G. West *The Ice Age in Britain* (Methuen), and for a study of rivers
see J. R. Lewin (ed.) *British rivers* (George Allen & Unwin). A useful account of
slopes and weathering is by R. J. Small & M. Clark *Weathering and slopes*
(Cambridge University Press). A good general account of the relations
between geology and scenery is provided by A. E. Trueman's *Geology and
scenery in England and Wales* (revised edition by J. B. Whittow & J. R. Hardy;
Pelican). In 1978/9 the *Geographical Magazine* had a series called 'Landform
Landmarks' in which some of the sites discussed in this book were described in
greater detail. Finally, the Geographical Association is now producing a series
of short booklets on classic landforms, and further details can be obtained
from The Geographical Association, 343 Fulwood Road, Sheffield. The series
is being produced in association with the society dedicated to the study of
landforms, The British Geomorphological Research Group (c/o Institute of
British Geographers, 1 Kensington Gore, London, SW7).

Index